Lecture Notes
in Computational Science
and Engineering

15

Editors

M. Griebel, Bonn

D. E. Keyes, Norfolk

R. M. Nieminen, Espoo

D. Roose, Leuven

T. Schlick, New York

Springer

Berlin
Heidelberg
New York
Barcelona
Hong Kong
London
Milan
Paris
Singapore
Tokyo

Andreas Frommer
Thomas Lippert
Björn Medeke
Klaus Schilling (Eds.)

Numerical Challenges in Lattice Quantum Chromodynamics

Joint Interdisciplinary Workshop
of John von Neumann Institute for Computing, Jülich,
and Institute of Applied Computer Science,
Wuppertal University, August 1999

With 32 Figures

 Springer

Editors

Andreas Frommer
Department of Mathematics
University of Wuppertal
42097 Wuppertal, Germany
e-mail:
frommer@math.uni-wuppertal.de

Björn Medeke
Department of Mathematics
University of Wuppertal
42097 Wuppertal, Germany
e-mail:
medeke@math.uni-wuppertal.de

Thomas Lippert
Department of Physics
University of Wuppertal
42097 Wuppertal, Germany
e-mail:
lippert@theorie.physik.uni-wuppertal.de

Klaus Schilling
John von Neumann Institute
for Computing
FZ-Jülich
52425 Jülich, Germany
e-mail:
schillin@theorie.physik.uni-wuppertal.de

Cataloging-in-Publication Data applied for

Die Deutsche Bibliothek - CIP-Einheitsaufnahme

Numerical challenges in lattice quantum chromodynamics : joint
interdisciplinary workshop of John von Neumann Institute for
Computing, Jülich, and Institute of Applied Computer Science,
Wuppertal University, August 1999 / Andreas Frommer ... (ed.). -
Berlin ; Heidelberg ; New York ; Barcelona ; Hong Kong ; London ;
Milan ; Paris ; Singapore ; Tokyo : Springer, 2000
  (Lecture notes in computational science and engineering ; 15)
  ISBN 3-540-67732-1

Mathematics Subject Classification (2000): 15-06, 15A15, 15A06

ISSN 1439-7358
ISBN 3-540-67732-1 Springer-Verlag Berlin Heidelberg New York

Springer-Verlag Berlin Heidelberg New York
a member of BertelsmannSpringer Science+Business Media GmbH

© Springer-Verlag Berlin Heidelberg 2000
Printed in Germany

Cover Design: Friedhelm Steinen-Broo, Estudio Calamar, Spain
Cover production: *design & production* GmbH, Heidelberg
Typeset by the authors using a Springer TEX macro package

Printed on acid-free paper   SPIN 10568848   46/3142/LK – 5 4 3 2 1 0

Preface

Lattice gauge theory is a fairly young research area in Theoretical Particle Physics. It is of great promise as it offers the framework for an ab-initio treatment of the nonperturbative features of strong interactions. Ever since its adolescence the simulation of quantum chromodynamics has attracted the interest of numerical analysts and there is growing interdisciplinary engagement between theoretical physicists and applied mathematicians to meet the grand challenges of this approach.

This volume contains contributions of the interdisciplinary workshop "Numerical Challenges in Lattice Quantum Chromodynamics" that the Institute of Applied Computer Science (IAI) at Wuppertal University together with the Von-Neumann-Institute-for-Computing (NIC) organized in August 1999. The purpose of the workshop was to offer a platform for the exchange of key ideas between lattice QCD and numerical analysis communities. In this spirit leading experts from both fields have put emphasis to transcend the barriers between the disciplines.

The meetings was focused on the following numerical bottleneck problems: A standard topic from the infancy of lattice QCD is the computation of Green's functions, the inverse of the Dirac operator. One has to solve huge sparse linear systems in the limit of small quark masses, corresponding to high condition numbers of the Dirac matrix. Closely related is the determination of flavor-singlet observables which came into focus during the last years. Stochastic estimates of such matrix traces suffer from severe fluctuations and therefore ask for extremely high statistics. At physical quark masses, high conditions numbers of the Dirac matrix are extremely demanding and ask for efficient preconditioning. Here, one would urgently wish to utilize multi-level algorithms which however are rendered inefficient due to random background gauge fields entering the coefficients of the lattice Dirac operator.

Recently, a very promising fermion discretization has been put forward. It respects a lattice symmetry which is closely related to chiral symmetry in the continuum and obeys the famous Ginsparg-Wilson relation. A particular scheme due to Neuberger, the overlap fermion formulation, requires the solution of linear systems including a matrix sign function. The computation of Green's functions in a standard approach becomes a real numerical bottleneck, being orders of magnitude more demanding than standard fermionic discretization schemes.

On future teracomputers, realistic simulations of QCD have to include three light dynamical quark flavors with non-degenerate masses. Multi-boson algorithms—approximating the inverse of the Dirac matrix by polynomial—in principle can deal with this situation. However, they require many auxiliary bosonic fields that slow down the stochastic sampling efficiency. On the other hand the alternative hybrid molecular dynamics algorithm is not exact and induces systematic errors.

We hope that the present volume reflects the informal crosstalk atmosphere of our workshop in dealing with these topics. Accordingly, the reader will find contributions from introductory to expert level.

We would like to mention that the Wuppertal meeting was a follow-up of a previous workshop organized by our colleagues of Kentucky University (Lexington) to stimulate synergism between lattice quantum chromodynamics and numerical analysis. We wish that these proceedings help strengthen existing and create new links between both communities.

Wuppertal, Jülich, April 2000   Andreas Frommer
                                                 Thomas Lippert
                                                 Björn Medeke
                                                 Klaus Schilling

Table of Contents

The Overlap Dirac Operator

Herbert Neuberger

Rutgers University
Department of Physics and Astronomy
Piscataway, NJ08855, USA

Abstract. This introductory presentation describes the Overlap Dirac Operator, why it could be useful in numerical QCD, and how it can be implemented.

1 Introduction

The objective of this talk is to introduce a certain matrix, the Overlap Dirac Operator. The ideas behind this construction are rather deep, having to do with the self-consistency of chiral gauge theories [1]. Chiral gauge theories are a class of quantum field theoretical models from which one draws the minimal standard model of particle physics. This model describes all phenomena at distances larger than about 10^{-18} meters, excluding gravity.

The Overlap Dirac Operator may be useful also in numerical studies of the strong interaction component of the minimal standard model, Quantum Chromo-Dynamics (QCD). QCD, in isolation, is a vector-like gauge theory, a simple combination of chiral gauge theories within which some of the mathematical complexities of the general class disappear. Nevertheless, the strong nonlinearity of QCD makes the quantitative evaluation of its most basic observable features possible only within large scale numerical Monte Carlo simulations. This workshop will deal mainly with numerical QCD, the subfield of lattice field theory which focuses on such simulations.

Therefore, the form of the Overlap Dirac Operator will be motivated heuristically, based on technical considerations of numerical QCD. The heuristic motivation will start from the continuum Euclidean Dirac operator, whose basic features of direct relevance to numerics will be reviewed. As alluded above, this is not the original way the Overlap Dirac Operator was found, but, for the focus of this workshop it is unnecessary to go through the entire construction. Rather, I shall start by reviewing an important technicality, the light quark bottleneck, which is a serious problem in contemporary traditional numerical QCD, argue that there is no fundamental reason why this bottleneck cannot be avoided and present the Overlap Dirac Operator as a potential solution[1]. The solution is not yet working efficiently because

---

[1] Other solutions than the Overlap Dirac Operator have been proposed for this specific problem; using the Overlap Dirac Operator is probably the most fundamentally motivated approach, but the cost may be high, and at the bottom line this consideration often takes precedence.

another bottleneck appears. However, the new bottleneck may be easier to circumvent than the old one.

An effort has been made to make this presentation accessible to applied mathematicians. Physicists practicing numerical QCD, in particular when the Overlap Dirac Operator is used, are facing issues that the expertise of specialists in numerical linear algebra could be of great help in resolving. These same physicists will likely be familiar with large portions of the presentation below, perhaps excepting the way the Overlap Dirac Operator is arrived at.

2   The Numerical Bottleneck of Light Quarks

The QCD Lagrangian with correct parameters should produce the observed ratio of masses $\frac{m_\pi}{m_\rho} = 0.18$. The relative smallness of this number reflects the relative small mass of two of the lightest quarks in Nature. This mass is not only small relatively to other quark masses, but, more importantly, relative to a basic scale of QCD itself, a mass parameter known as Λ_{QCD}. Theoretically, one has very good reasons to expect to be able to smoothly extrapolate in the light quark masses from, say, $\frac{m_\pi}{m_\rho} = 0.25$ to the smaller physical value. Actually, even the functional form of these extrapolations is known. However, it is not likely that the extrapolation will go smoothly through the value $\frac{m_\pi}{m_\rho} = 0.5$. As long as $m_\pi m_\rho \geq 0.5$ the decay $\rho \to 2\pi$ is energetically forbidden but once this threshold is passed the decay becomes allowed. This provides physical reasons to expect a "bumpy" behavior[2] in the extrapolation to light quark masses in the region where $\frac{m_\pi}{m_\rho} \sim 0.5$ and there are no first principle based theoretical expressions parameterizing this "bumpy" region. Current large scale numerical work happens to be close to the threshold $\frac{m_\pi}{m_\rho} = 0.5$, and there are numerical difficulties obstructing straightforward attempts to go significantly lower.

In numerical work one uses a lattice whose spacing a is measured in inverse mass units. Thus any number of the type mass $\times$ a is a pure number, of the kind a computer can handle. The threshold ratio is obtained, with current simulation parameters, for $m_\pi a \sim 0.1$ and for quark mass m_q given by $m_q a \sim 0.05$. Nothing is lost by setting $a = 1$; dimensional considerations can

---

[2] Some quantities may be insensitive to the $\rho \to 2\pi$ threshold effects, typically for kinematic reasons. Also, there is an additional approximation which is very often employed (the so called "valence", or "quenched" approximation) which eliminates decay effects on the vacuum of the theory. On the other hand, one often relies on an effective low energy description of QCD, in which the Lagrangian is replaced by an "effective Lagrangian" involving only π's and some other light associated particles. The existence of a particle like the ρ and its connection to the 2π state is then encoded in the coefficients of the "effective Lagrangian" and will affect the accuracy of chiral perturbation theory at moderate energies. Thus, physical observables related to "weak matrix elements" are an example where one might suspect sensitivity to threshold effects.

always restore the right power of a. Physical considerations imply $m_\pi \propto \sqrt{m_q}$ for small m_q. On the other hand, m_ρ stays finite (basically of the order of Λ_{QCD}) at $m_q = 0$ and consequently its numerical dependence on m_q for small m_q is weaker. So, in order to reduce $\frac{m_\pi}{m_\rho}$ by a factor of 2 as we would like, we have to reduce the light quark masses by about a factor of 4. The bottom line is that we would like to leave all other simulation parameters the same, only change the quark mass to something like $m_q a = 0.01$.

Simple arguments indicate that we should be able to do that. The quark mass enters as an important parameter in the lattice Dirac Operator, D. D and $D^\dagger D$ are matrices that need to be inverted on vectors in the most inner procedure of the simulations. In traditional simulations D is a sparse matrix; full storage is out of the question, as the dimension of D is too large. The quark mass directly affects the condition number of D, so its value can easily become the bottleneck of the entire simulation. But, we could easily imagine being lucky enough to manage the needed value of $m_q a = 0.01$: D is a discretized first order partial differential operator and, since space time is four dimensional one can easily think of D obeying $\|D\| \le 4$ as a uniform upper bound. On the other hand, m_q enters additively, with unit coefficient and, based on the continuum formula one would expect that D is approximately anti-hermitian at zero quark mass. Therefore, we expect $\|D^\dagger D\| \ge m_q^2$ and, when we invert $D^\dagger D$, a condition number of the order 10^5, which should allow the evaluation of $\frac{1}{D^\dagger D}\psi$ in something like 500-1000 iterations.

Up to relatively short lived jerks, computational power is projected to increase by Moore's Law for another decade, so we can expect an enhancement by a factor of 250 at fixed power dissipation by 2010. Based on the accumulated numerical experience to date, we can reasonably expect that, by then, we would be able to carry out our computations with relative ease.

However, the sad fact is that theoretical issues have apparently made it impossible to find a simple enough D which would obey the rather plausible lower bound $\|D^\dagger D\| \ge m_q^2$. This was a key assumption I made before in order to come up with the condition number of order 10^5. What really happens in the traditional approach[3] is that $D^\dagger + D$ is not small and to get $\frac{m_\pi}{m_\rho} \sim .3$ one is forced into a regime where the lowest eigenvalue of $D^\dagger D$ quite often fluctuates as low as 10^{-8}. This makes the condition number go to 10^{10} and puts the $\frac{m_\pi}{m_\rho} \sim .3$ mass ratio out of numerical reach.

The above numerical problem reflects a difficulty of principle associated with the $m_q = 0$, or massless, case. In the continuum model, $m_q = 0$ is associated with an extra symmetry, chirality. Technically, chirality ensures that in the continuum the massive Dirac operator $D + m_q$ (where I reserved D for the massless case) indeed obeys $\|D^\dagger D\| \ge m_q^2$. But, chirality cannot be preserved on the lattice in the same way as some other symmetries can. For twenty years it was thought that an acceptable lattice version of D which

---

[3] To be specific, the approach employing Wilson fermions; there also exists a traditional "staggered fermion" alternative, but it suffers from other problems.

also preserves chirality does not exist. The good news is that a series of developments, started in 92 by Kaplan and by Frolov and Slavnov [2] and built on by Narayanan and myself, have produced, in 97, the Overlap Dirac Operator D_o on the lattice, which, effectively has the extra symmetry. While D_o is easy to write down, it is not easy to implement because it no longer is sparse. There nevertheless is some hope, because, in the most common implementation of the Overlap ideas, D_o is what probably would be the next best thing: up to trivially implemented factors, it is a function (as an operator) of a sparse matrix[4]. Unfortunately, the function has a discontinuity, so its implementation is costly. The numerical developments in this subfield are relatively fresh and substantial progress has been achieved but we are not yet adequately tooled for full "production runs". Still, since the time we have had to address the problem is of the order of a year or two I think further substantial progress is likely.

3   Dirac Operator Basics

The Dirac Operator is a very important object in relativistic Field Theory. Among other things, it predicted the existence of the electron anti-particle, the positron. All of known matter is governed by the Dirac equation.

3.1   Continuum

The Dirac Operator is defined in the continuum (Euclidean space) by

$$D_c = \sum_\mu \gamma_\mu (\partial_\mu + i A_\mu) \; , \tag{1}$$

where,

$$\gamma_\mu^\dagger = \gamma_\mu; \quad \mu = 1, 2, 3, 4; \quad \{\gamma_\mu, \gamma_\nu\} \equiv \gamma_\mu \gamma_\nu + \gamma_\nu \gamma_\mu = 2\delta_{\mu\nu} \; . \tag{2}$$

We also have,

$$A_\mu = A_\mu^\dagger; \quad tr A_\mu = 0; \quad A_\mu \text{ are 3 by 3 matrices.} \tag{3}$$

The notation ∂_μ indicates $\partial/\partial x_\mu$. A_μ is x-dependent, but the γ_μ are constant four by four matrices and operate in a separate space. Thus, D_c is a twelve by twelve matrix whose entries are first order partial differential operators on a four dimensional Euclidean space, henceforth assumed to be a flat four torus. The massive Dirac Operator is $m_c + D_c$.

The main properties of the Dirac Operator are:

---

[4] The additional "trivial" factors have a highly non-trivial numerical impact: because of them the *inverse* of D_o no longer has a simple expression in terms of a function of a sparse matrix.

1. For $A_\mu = 0$ one can set by Fourier transform $\partial_\mu = ip_\mu$ giving $D_c = i\gamma_\mu p_\mu$. This produces $D_c^2 = -\sum_\mu p_\mu^2$ from the algebra of the γ-matrices. The important consequence this has is that $D_c = 0$ iff $p_\mu = 0$ for all μ.
2. Define $\gamma_5 \equiv \gamma_1\gamma_2\gamma_3\gamma_4$ which implies $\gamma_5^2 = 1$ and $\gamma_5\gamma_\mu + \gamma_\mu\gamma_5 = 0$. The property of γ_5-hermiticity is that D_c is hermitian under an indefinite inner product defined by γ_5, or, equivalently, using the standard L_2 inner product, we have for any backgrounds A_μ and any mass parameter m_c

$$\gamma_5(D_c + m_c)\gamma_5 = (D_c + m_c)^\dagger \ . \tag{4}$$

Thus, γ_5-hermiticity is a property of the massive Dirac Operator.
3. At zero mass the Dirac Operator is anti-hermitian $D_c^\dagger = -D_c$ This property, in conjunction with γ_5-hermiticity implies the symmetry property of chirality: $\gamma_5 D_c \gamma_5 = -D_c$ usually written as $\{D_c, \gamma_5\} = 0$. Traditionally, anti-hermiticity would be viewed as a separate and trivial property and chirality as a symmetry. Then, γ_5 hermiticity would be a consequence holding not only for the massless Dirac Operator, but also for non-zero mass. But, for better comparison to the lattice situation it is somewhat better to exchange cause and consequence, as done here.

3.2   Lattice

Numerical QCD spends most of the machine cycles inverting a lattice version of the massive Dirac Operator. This is how the continuum Dirac operator is discretized:

The continuum x is replaced by a discrete lattice site x on a torus. The matrix valued functions A_μ are replaced by unitary matrices $U_\mu(x)$ of unit determinant. $U_\mu(x)$ is associated with the link l going from x into the positive μ-direction, $\hat{\mu}$.

The connection to the continuum is as follows: Assume that the functions[5] $A_\mu(x)$ are given. Then,

$$U_\mu = \lim_{N\to\infty} \left[e^{iaA_\mu(x)/N} e^{iaA_\mu(x+a)/N} e^{iaA_\mu(x+2a)/N} \ldots e^{iaA_\mu(x+(N-1)a))/N} \right]$$
$$\equiv P\exp[i\int_l dx_\mu A_\mu(x)] \ \text{(the symbol P denotes "path ordering")} \ . \tag{5}$$

There are two sets of basic unitaries acting on twelve component vectors $\psi(x)$:

1. The point wise acting γ_μ's and

---

[5] Actually, the $A_\mu(x)$ aren't really functions, rather they make up a one form $\sum_\mu A_\mu(x)dx_\mu$ which is a connection on a possibly nontrivial bundle with structure group $SU(3)$ over the four-torus. This complication is important for the case of massless quarks, but shall be mostly ignored in the following.

2. the directional parallel transporters T_μ, defined by:

$$T_\mu(\psi)(x) = U_\mu(x)\psi(x + \hat{\mu}) \ . \tag{6}$$

There also is a third class of unitaries carrying out gauge transformations. Gauge transformations are characterized by a collection of $g(x) \in SU(3)$ and act on ψ pointwise. The action is represented by the unitary $G(g)$, so that $(G(g)\psi)(x) = g(x)\psi(x)$. Probably the most important property of the T_μ operators is that they are "gauge covariant",

$$G(g)T_\mu(U)G^\dagger(g) = T_\mu(U^g) \ , \tag{7}$$

where,

$$U^g_\mu(x) = g(x)U_\mu(x)g^\dagger(x + \hat{\mu}) \ . \tag{8}$$

The variables $U_\mu(x)$ are stochastic, distributed according to a probability density that is invariant under $U \to U^g$ for any g.

The lattice replacement of the massive continuum Dirac Operator, $D(m)$, is an element in the algebra generated by T_μ, $T_\mu^\dagger$, γ_μ. Thus, $D(m)$ is gauge covariant. For $U_\mu(x) = 1$ the T_μ become commuting shift operators. One can preserve several crucial properties of the continuum Dirac Operator by choosing a $D(m)$ which satisfies:

1. Hypercubic symmetry. This discrete symmetry group consists of 24 permutations scrambling the μ indices (which leave $D(m)$ unchanged) combined with 16 reflections made out of the following four elementary operations:

$$\gamma_5\gamma_\mu D(T_\mu \longleftrightarrow T_\mu^\dagger)(m)\gamma_\mu\gamma_5 = D(m) \ . \tag{9}$$

2. Correct low momentum dependence. When the commutators $[T_\mu, T_\nu] = 0$ one can simultaneously diagonalize the T_μ unitaries, with eigenvalues $e^{i\theta^k_\mu}$. For small θ angles, one gets $D(m) \sim m + i\sum_\mu \gamma_\mu\theta_\mu + O(\theta^2)$. One also needs to require that for zero mass $(m = 0)$ $D^\dagger D$ be bounded away from zero for all θ which are outside a neighborhood R of zero.

3. Locality and smoothness in the gauge field variables. One can also require, at least for gauge backgrounds with small enough (in norm) commutators $[T_\mu, T_\nu]$, that D be a convergent series in the T_μ.

The simplest solution to the above requirements is the Wilson Dirac Operator, D_W. It is the sparsest possible realization of the above. Fixing a particular free parameter (usually denoted by r) to its preferred value ($r = 1$) D_W can be written as:

$$D_W = m + 4 - \sum_\mu V_\mu; \quad V_\mu^\dagger V_\mu = 1; \quad V_\mu = \frac{1 - \gamma_\mu}{2}T_\mu + \frac{1 + \gamma_\mu}{2}T_\mu^\dagger \ . \tag{10}$$

For $m > 0$ clearly $\|D_W\| \geq m$, but to get the physical quark mass to zero, $m_q^{\text{phys}} = 0$, one needs to take into account an additive renormalization induced by the fluctuations in the gauge background $U_\mu(x)$: m must be chosen as $m = m_c(\beta) < 0$ where β is the lattice version of the gauge coupling constant governing the fluctuations of the background. The need to use a negative m on the lattice opens the door for very small eigenvalues for $D_W^\dagger D_W$ and thus for terrible condition numbers. To get a small but nonzero physical quark mass one should choose $m = m_c(\beta) + \Delta m$, $\Delta m > 0$ but the numbers are such that one always ends up with an overall negative m.

3.3   A Simplified "No-Go Theorem".

These problems would go away if we had chirality, because it would single out the $m = 0$ values as special and remove the additive renormalization. But, this cannot be done. There are rather deep reasons for why this cannot be done, and several versions of "no-go theorems" exist[3]. Here, I shall present only a very simple version, namely, one cannot supplement the above requirements of $D(m)$ (fulfilled by D_W) with the requirement $\gamma_5 \tilde{D} \gamma_5 = -\tilde{D}$ at $m = 0$. The proof is very simple: Take the particular case where the T_μ commute. Compounding four elementary reflections we get $\gamma_5 \tilde{D}(\theta) \gamma_5 = \tilde{D}(-\theta) = -\tilde{D}(\theta)$. This shows that any one of the sixteen solutions $\theta_\mu = 0, \pi$ has $\tilde{D}(\theta^*) = 0$, since, by periodicity $\pi = -\pi$. This violation of one of the original requirements amounts to a drastic multiplication of the number of Dirac particles: instead of one we end up with sixteen !

4   Overlap Dirac Operator

Any regularization deemphasizes the dynamics of short length/time scales. Lattice regularizations completely remove arbitrarily high momenta by compactifying momentum space. Thus, the spectrum of a lattice regularized Dirac operator lives on a compact space. When looking for a way around the chirality "no-go theorem" a possible line of attack is to prepare ahead of time for the regularization step by shifting the focus to an operator which already has a compact spectrum in the continuum.

Based on the anti-hermiticity of D_c (which we argued before can be viewed as a fundamental property, one that in conjunction with the other fundamental property of γ_5-hermiticity produces chirality as a consequence) we can quite naturally define an associated unitary operator V_c using the Cayley transform:

$$V_c = \frac{D_c - \Lambda_c}{D_c + \Lambda_c} \ . \tag{11}$$

V_c is not only unitary, but also γ_5-hermitian, inheriting the property from D_c. In other words, V_c has the two crucial properties of D_c and these two properties would imply chirality for D_c defined in terms of V_c by inverting the Cayley transform (the inverse is another Cayley transform). Also, one has to note that the "no-go theorem" does not prohibit a lattice version of V_c, V, that satisfies both requirements of γ_5-hermiticity and unitarity. However, the D one would construct from such a V would need to violate something - the natural choice is that D be non-local. Still, nothing seems to say that V itself has to be nonlocal:

$$D = \frac{1+V}{1-V} \ . \tag{12}$$

The non-locality in D, mandated by the "no-go theorem", could reflect merely the existence of unit eigenvalues to V.

The next question is then, suppose we have a lattice V; since D is non-local, how can we hope to make progress ? The answer is almost trivial if we go back to continuum: we only care about the spectral part of D_c which is below some cutoff Λ_c. So, we are allowed to expand the Cayley transform:

$$V_c = -1 + 2\frac{D_c}{\Lambda_c} + O\left(\frac{D_c}{\Lambda_c}\right)^2 \ . \tag{13}$$

Therefore,

$$\frac{D_c}{\Lambda_c} = \frac{1+V_c}{2} \quad \text{plus unphysical corrections.} \tag{14}$$

If the lattice-V is local there is no locality problem with the lattice-D_o being given by

$$D_o = \frac{1+V}{2} \ . \tag{15}$$

Now, in agreement with the "no-go theorem", we have lost exact anti-hermiticity, and, consequently exact, mathematical, chirality. However, the loss of chirality can be made inconsequential, unlike in the traditional Wilson formulation of fermions. As a first sign of this let us check that adding a mass term produces a lower bound similar to the continuum and thus would protect our precious condition number. It is easy to prove that:

$$\left(\frac{m_c}{\Lambda_c} + \frac{1+V_c}{2}\right)^\dagger \left(\frac{m_c}{\Lambda_c} + \frac{1+V_c}{2}\right) \geq \min\left[\left(\frac{m_c}{\Lambda_c}\right)^2, \left(\frac{m_c}{\Lambda_c} + 1\right)^2\right] \ . \tag{16}$$

So, we have an expression bounded away from zero as long as $\frac{m_c}{\Lambda_c} \neq 0, -1$. Another way to see that we have as much chirality as we need is to calculate the anticommutator of D_o with γ_5, which would be zero in the continuum, when D_c is used. Actually, in the language of path integrals, we do not really need the Dirac Operator itself: rather we need its inverse and its determinant.

So, it is closer to physics to look at the inverse of the Overlap Dirac Operator D_o. D_o^{-1} obeys

$$\gamma_5 D_o^{-1} + D_o^{-1} \gamma_5 = 2\gamma_5 \ . \tag{17}$$

The same equation would also hold in the continuum if we replaced D_c^{-1} by $\frac{2}{1+V_c}$. There is little doubt that such a replacement in the continuum has no physical consequence; after all it only changes D_c^{-1} additively, by -1, a Dirac delta-function in space-time. While on the lattice we don't have a local and exactly chiral D, we do have a $D_o = \frac{1+V}{2}$ and we now see that *its* inverse is good enough. In a brilliant paper [4][6], published almost twenty years ago, Ginsparg and Wilson suggested this relaxation of the chirality condition as a way around the "no-go theorem". What stopped progress was the failure of these authors to also produce an explicit formula for D_o. Their failure, apparently, was taken so seriously, that nobody tried to find an explicit D_o, and the entire idea fell into oblivion for an embarrassingly long time. Nobody even made the relatively straightforward observation that the search for D_o was algebraically equivalent to a search for a unitary, γ_5-hermitian operator V. Moreover, it seems to have gone under-appreciated that the problem was not so much in satisfying the algebraic constraint of the relaxed form of the anticommutator (the GW relation), but, in simultaneously maintaining gauge covariance, discrete lattice symmetries, and correct low momentum behavior, without extra particles.

4.1   Definition and Basic Properties

To guess a lattice formula for V, first notice that, in the continuum, the operator $\epsilon_c = \gamma_5 V_c$ is a reflection: $\epsilon_c^2 = 1$ and $\epsilon_c^\dagger = \epsilon_c$. Also, V_c, and hence ϵ_c, are expressed as a ratio of *massive* Dirac operators. But, we know that there is no difficulty in representing on the lattice the massive Dirac Operator. The simplest way to do this is to use the Wilson Dirac Operator, D_W. The latter obeys γ_5-hermiticity, so $H_W = \gamma_5 D_W$ is hermitian. It is then very natural to try

$$\epsilon = \text{sign}(H_W) \ . \tag{18}$$

This is not the entire story - we still have one parameter at our disposal, the lattice mass, m. When we look at the continuum definition of V_c we observe that it can be rewritten as follows:

$$V_c = \frac{D_c - \Lambda_c}{\sqrt{-D_c^2 + \Lambda_c^2}} \frac{\sqrt{-D_c^2 + \Lambda_c^2}}{D_c + \Lambda_c} \ . \tag{19}$$

---

[6] Clearly, when going to the lattice, the continuum delta-function can be replaced by a Kroneker delta-function or by, say, a narrow Gaussian. This gives a certain amount of freedom which I shall ignore below. In the lattice formulation of Ginsparg and Wilson this freedom is made explicit by the introduction of a local operator R.

So V_c is given by the product of two unitary operators, one being the unitary factor of the Dirac Operator with a large negative mass and the other the conjugate of the unitary factor of the Dirac Operator with a large positive mass. Formally, the unitary factors are equal to each other up to sign in the continuum and one has:

$$V_c = -U_c^2 = -\left(\frac{D_c - \Lambda_c}{\sqrt{-D_c^2 + \Lambda_c^2}}\right)^2 ; \quad -D_c^2 + \Lambda_c^2 = (D_c - \Lambda_c)^\dagger (D_c - \Lambda_c) .$$

$$(20)$$

But, on the lattice, a local unitary operator U replacing U_c cannot exist; if it did, $D = \frac{1}{2}(U^\dagger - U)$ would violate the "no-go theorem". So, on the lattice we can find a V, but there is no local square root of $-V$; this observation is rather deep, and related to anomalies. We need to treat the two unitary factors in the continuum expression for V_c differently on the lattice. Actually, the factor with a positive mass sign can be replaced by unity by taking the limit $m \to \infty$, something one can do only on the lattice, where the T_μ operators are bounded. The factor representing negative mass however cannot be so simplified because the negative mass argument m is restricted to a finite interval $(-2 < m < 0)$ to avoid extra particles. Thus, on the lattice we are led to

$$V = D_W(m) \frac{1}{\sqrt{D_W^\dagger(m) D_W(m)}} \quad \text{with} \quad -2 < m < 0 . \quad (21)$$

This formula is equivalent to the one for ϵ above. The reason that we get exactly massless quarks is that no fine tuning is needed for V to have eigenvalues very close to -1. Indeed, $D_W(m) = D_W(0) + m$ and, in the case $[T_\mu, T_\nu] = 0$, $D_W(0)$ is easily seen to have very small eigenvalues. There $D_W(m)$ is dominated by the negative mass m. It is unimportant what the exact value of m is, only its sign matters. Even when $[T_\mu, T_\nu] \neq 0$, and m is additively renormalized, as long as the effective m stays negative, we shall have exactly massless quarks as evidenced by the eigenvalues of V close to -1 potentially producing long range correlations in D_o^{-1}.

All non-real eigenvalues of V are paired: $V\psi = e^{iv}\psi \Rightarrow V\gamma_5\psi = e^{-iv}\gamma_5\psi$. But, one important property of the continuum V_c is that it has an unpaired single -1 eigenvalue in a certain class of topologically nontrivial backgrounds. This is a characteristic of massless quarks and has to be reproduced on the lattice. There V is even dimensional, so a single exact $V = -1$ state implies the existence of another $V = +1$ state. In addition, it is clear that as a result of $|m| < 2$, there are states where $D_W(0)$ dominates over m and these states will have eigenvalues far off the real axis. We conclude that the spectrum of V can cover the entire complex unit circle. The spectrum of U_c though, covers only half the complex unit circle. This is a reflection of V being a lattice version of $-U_c^2$ rather than U_c itself, as one might think naively.

It was mentioned already that all one needs in the path integral are formulae for the inverse of the Dirac Operator and for its determinant. It is clear that once one has an acceptable V one can form not only the local, but not strictly chiral operator D_o but also a non-local chiral associate $D_o^\chi = \frac{1+V}{1-V}$. In the determinant one must use D_o and cannot use D_o^χ. Otherwise, the extra zeros of the determinant coming from the $V = 1$ states leave an indelible effect in the continuum limit. These zeros directly reflect the non-locality of D_o^χ. However, the inverses of D_o can be replaced by inverses of D_o^χ which has useful practical implications. In the language of Feynman graphs, the determinant is represented by closed internal fermion loops, and on those we must use D_o^{-1}, but the inverses come from external fermions lines, and on those we can just as well use $D_o^\chi{}^{-1} = D_o^{-1} - 1$. This is consistent because the -1 term can be interpreted as coming from an auxiliary fermionic variable which contributes only a unit multiplicative factor to the fermionic determinant.

To complete the logic of the story let me discuss the introduction of a finite quark mass parameter in the context of the Overlap Dirac Operator. To be sure there is no additive mass renormalization, one wishes to preserve the continuum property that $Tr\frac{1}{D_c+m_c}$ is odd in m_c; this property[7] singles out the point $m_c = 0$. This property is a direct consequence of $\gamma_5 D_c + D_c\gamma_5 = 0$. It is easy to see that for an external fermion line the lattice version would be

$$Tr\,\frac{e^\rho - V}{e^\rho + V} = Tr\left(\frac{2}{1 + e^{-\rho}V} - 1\right); \qquad \rho > 0; \tag{22}$$

with

$$m_q = z\tanh\frac{\rho}{2}; \qquad \left(\frac{1+V}{2}\right)_{T_\mu = \exp(i\theta_\mu)} \longrightarrow \frac{i}{z}\sum_\mu \gamma_\mu\theta_\mu + O(\theta_\mu^2)\ . \tag{23}$$

The factor z is somewhere between 1 and 2. One now easily proves that the condition number of the matrix $\frac{e^\rho+V}{e^\rho-V}$ is $\frac{1}{\tanh(\frac{\rho}{2})} = \frac{z}{m_q}$. Therefore, we should be able to get to the quark masses we need. However, the matrix ϵ isn't sparse, and the evaluation of its action will be time consuming. Still, it is a function of a sparse matrix H_W, so employing the Overlap Dirac Operator in numerical QCD is not ruled out *ab initio*.

4.2   Implementation by Rational Approximants

The operator ϵ is unambiguously defined only for matrices H_W that have no zero eigenvalues. This is not a restriction in itself, because the variables $U_\mu(x)$ in H_W are stochastic and there is no symmetry protecting zero eigenvalues of H_W. So, the probability to encounter an exact zero of H_W is zero. Moreover, in the continuum limit H_W will have a relatively large gap around zero, so

---

[7] A less careful introduction of non-zero quark mass in the Overlap Dirac Operator requires extra unnecessary subtractions - see [6].

when we are close enough to the continuum we need the sign function only over the range of arguments $[-a, -b] \cup [b, a]$, where $a \sim 8$ and b should be something like 0.1 or 0.5.

Over the above range it is relatively easy to approximate the sign function by simpler functions. Also, for b uniformly (in the background) bounded away from zero, the operator $\frac{1}{\sqrt{D_W^\dagger D_W}}$ is a convergent series in H_W. This ensures that the non-sparse D_o is nevertheless sufficiently local to be an acceptable approximation to a continuum differential operator.

Mimicking the way transcendental functions are typically implemented in computers, we are led to try a rational approximation [7]. So, we are looking for a series of functions $\varepsilon_n(x)$ which, for any fixed $x \neq 0$, obeys $\lim_{n\to\infty} \varepsilon_n(x) = \text{sign}(x)$. For each finite n, $\varepsilon_n(x)$ is a ratio of two polynomials. A sequence with many good properties has been in use by applied mathematicians [5] who were interested in the sign function because of the role it plays in control theory. The sequence is given by:

$$\varepsilon_n(x) = \frac{(1+x)^{2n} - (1-x)^{2n}}{(1+x)^{2n} + (1-x)^{2n}} = \begin{cases} |x| < 1 \ \tanh[2n \tanh^{-1}(x)] \\ |x| > 1 \ \tanh[2n \tanh^{-1}(x^{-1})] \\ |x| = 1 \ x \end{cases} \quad (24)$$

So long $|x|$ is sufficiently far from zero a large enough n can provide an approximation to the sign function good to machine accuracy. The approximants $\varepsilon_n(x)$ are smooth and maintain some properties of the sign function:

$$\varepsilon_n(x) = -\varepsilon_n(-x) = \varepsilon_n(\frac{1}{x}); \qquad |\varepsilon_n(x)| \leq 1; \qquad \varepsilon_n(\pm 1) = \pm 1 . \quad (25)$$

To implement $\varepsilon_n(x)$ we decompose $f(x^2) = \varepsilon_n(x)/x$ in simple pole terms:

$$\varepsilon_n(x) = \frac{x}{n} \sum_{s=1}^{n} \frac{1}{x^2 \cos^2 \frac{\pi}{2n}(s - \frac{1}{2}) + \sin^2 \frac{\pi}{2n}(s - \frac{1}{2})} . \quad (26)$$

One can choose an appropriate scaling parameter $\lambda > 0$ and approximate the sign function of H_W by $\varepsilon_n(\lambda H_W)$. The action of this approximated ϵ on a vector ψ can be evaluated by a multi-shift Conjugate Gradient inversion algorithm The multi-shift trick reduces the computational cost of evaluating the action of each one of the terms in the pole expansion to the cost of evaluating one single inversion, the most time consuming one, which clearly is

$$\frac{1}{\lambda^2 H_W^2 + \tan^2 \frac{\pi}{4n}} \psi . \quad (27)$$

Memory requirements are linear in n, since one needs to store a few vectors for each pole term. In practice this may be a problem, not as much that the entire memory of the machine would be exceeded, but rather that one would find oneself exceeding the level 2 cache. Level 2 cache misses can lead to substantial performance loss. Thus, it is useful to consider an alternative to the standard implementation of the multi-shift trick. Usually, in applications of the multi-shift trick, one is really interested in the individual vectors, but here we only want their weighted sum. It is quite easy to figure out a way to store only a few vectors, and get the sum, but a double pass over the basic Conjugate Gradient procedure is now required. Although the number of floating point operations is doubled, because of reduced cache miss penalties, performance is not necessarily adversely impacted and can actually increase [9].

Other promising methods to implement D_o have been developed and will be hopefully discussed here later [10]. In the implementation employed in [6] the sign function was approximated by a very high order polynomial.

4.3 A New Bottleneck and Projectors

Having removed one apparent obstacle, namely the potentially large memory requirements, we turn to a much more substantial obstacle, namely the condition number of H_W^2, κ. κ determines both the number of Conjugate Gradient iterations needed to evaluate the inverse and also the minimal required n for given expected accuracy δ, where

$$\|\varepsilon_n(\lambda H_W) - \text{sign}(H_W)\| < \delta \ . \tag{28}$$

The optimal choice of λ is easily found because of the inversion symmetry of $\varepsilon_n(x)$: $\lambda h_{\min} = \frac{1}{\lambda h_{\max}}$ where $h_{\min, \max}$ are the square roots of the minimal and maximal eigenvalues of H_W^2. Thus, $\lambda = \frac{1}{\sqrt{h_{\min} h_{\max}}}$ so the range over which the sign function is needed is

$$\frac{1}{\kappa^{\frac{1}{4}}} < |x| < \kappa^{\frac{1}{4}} \ . \tag{29}$$

For $\varepsilon_n(x)$ to be an acceptable approximation we need then $n \approx \frac{1}{4}\kappa^{\frac{1}{4}}|\log(\delta/2)|$. For example, for $\kappa \sim 10^4$ and single precision $n \sim 50$ is safe; for double precision, the needed n doubles. On the other hand the number of Conjugate Gradient iterations will go as $\kappa^{\frac{1}{2}}$. Note that the lowest eigenvalue of $\lambda^2 H_W^2$ and the minimal pole shift $\frac{\pi^2}{16n^2}$ are of the same order when $n \sim \kappa^{\frac{1}{4}}$.

Comparing to traditional simulations, where κ is again the source of all problems, it is important to note the new feature that now the parameter m in H_W is taken in a different regime from before. In traditional simulations the parameter m is adjusted so that the physical quark mass be small. This precisely means that the parameter is chosen so that κ be large. Because of fluctuations, κ becomes often much larger than one would have expected, and that is the problem faced by traditional simulations discussed earlier. Here the parameter m, subject to the limitation $-2 < m < 0$, can be chosen so that κ be as small as possible. Fluctuations can still cause problems and sometimes make κ large, but, in principle, the fluctuations are around a small κ value, not a large one. In practice however the situation is somewhat marginal: coarse lattices produce too much fluctuations even for the Overlap Dirac Operator. But, finer lattices are manageable. Still, even on finer lattices one needs to split the domain over which the sign function is approximated into a neighborhood around zero and the rest. The neighborhood around zero contains of the order of ten eigenvalues, and by computing the exact projector on the corresponding eigenspace, the sign function is exactly evaluated there, leaving only the more manageable part of the spectrum to be dealt with by the rational approximation [11]. This is quite time consuming, and constitutes the new bottleneck we are facing. Variations of the probability distribution of the background and slight modifications of H_W would not affect the continuum limit but would probably help ameliorate this problem.

4.4   Avoiding Nested Conjugate Gradient Procedures

Another obvious drawback of using the Overlap Dirac Operator in numerical simulations is that we eventually need to compute vectors of the form $\frac{2}{1+V}\psi$ for a given ψ. This requires another inversion and we end up with a two level nested Conjugate Gradient procedure, whereas in the traditional simulations we only had one. The outer Conjugate Gradient is going to be better conditioned (for light but still massive quarks) than in the traditional approach. So, we are facing a tradeoff issue, one that has not been resolved yet. The two levels of Conjugate Gradients are presently dealt with separately, but clearly, eventually, potential gains will be obtained from the fact that the inner procedure does not have to be run to high accuracy for the initial steps in the outer procedure.

There exists another trick [12] to simulate the rational $\varepsilon_n(\lambda H_W)$ which does away with the need of employing a nested Conjugate Gradient altogether, but at the expense of memory usage linear in n. Basically, the point is to find a Conjugate Gradient algorithm operating in an enlarged space and producing the vector $\frac{2}{1+V}\psi$ in one blow. This trick is based on two observations:

1. Any rational approximation can be written as a truncated continued fraction.
2. Any truncated continued fraction can be represented by a Gaussian Path Integral of a fermionic system living on a chain of length n, where n is the depth of the truncated continued fraction.

This method is applicable to any rational approximation. Below is one way to apply it to $\varepsilon_n(H_W)$, where the scale factor is absorbed in H_W for notational simplicity.

First, the rational approximation has to be written in the form of a continued fraction with entries preferably linear in H_W. I start from a formula that goes as far back as Euler, and subsequently use the invariance under inversion of x to move the x factors around, so that the entries become linear in x.

$$\varepsilon_n(x) = \cfrac{2nx}{1 + \cfrac{(4n^2 - 1)x^2}{3 + \cfrac{(4n^2 - 4)x^2}{5 + \ldots \cfrac{\ddots}{4n - 3 + \cfrac{[4n^2 - (2n-1)^2]x^2}{4n - 1}}}}} \tag{30}$$

Now, with the help of extra fields, I write a Gaussian path integral which induces the desired action between a chosen subset of fields:

$$\int d\bar{\phi}_1 d\phi_1 d\bar{\phi}_2 d\phi_2 \ldots d\bar{\phi}_n d\phi_n e^{S_*} = (\det H_W)^{2n} e^{-\bar{\psi}(\gamma_5 + \varepsilon_n(H_W))\psi} \ . \tag{31}$$

The quadratic action S_* couples the extended fermionic fields $\bar{\chi}, \chi$:

$$\bar{\chi} = \left(\bar{\psi}\ \bar{\phi}_1\ \ldots\ \bar{\phi}_{2n} \right), \qquad \chi = \begin{pmatrix} \psi \\ \phi_1 \\ \vdots \\ \phi_{2n} \end{pmatrix} . \tag{32}$$

$S_* = \bar{\chi}\mathbf{H}\chi$, where the new kernel, $\mathbf{H}$, has the following block structure:

$$\mathbf{H} = \begin{pmatrix} -\gamma_5 & \sqrt{\alpha_0} & 0 & \cdots & \cdots & 0 \\ \sqrt{\alpha_0} & H_W & \sqrt{\alpha_1} & \cdots & \cdots & 0 \\ 0 & \sqrt{\alpha_1} & -H_W & \cdots & \cdots & 0 \\ \cdots & \cdots & \cdots & \ddots & \cdots & 0 \\ \cdots & \cdots & \cdots & \cdots & H_W & \sqrt{\alpha_{2n-1}} \\ \cdots & \cdots & \cdots & \cdots & \sqrt{\alpha_{2n-1}} & -H_W \end{pmatrix} . \tag{33}$$

The numerical coefficients α are given below:

$$\alpha_0 = 2n, \ \alpha_j = \frac{(2n - j)(2n + j)}{(2j - 1)(2j + 1)}, \ j = 1, 2, \ldots \tag{34}$$

The hope is that the condition number of $\mathbf{H}$ will be manageable. The basic point is that up to a scalar factor, the (1,1) diagonal block of the inverse $\mathbf{H}^{-1}$ is equal to $\frac{1}{1+V}\gamma_5$. $\mathbf{H}$ is sparse and the evaluation of $\mathbf{H}^{-1}\chi$ requires one single Conjugate Gradient procedure, albeit one acting on vectors $2n$ times longer than needed. Although I used Path Integrals to get the relation between $\mathbf{H}$ and the Overlap Dirac Operator, there is nothing more to the derivation than ordinary Linear Algebra, and Path Integrals could have been bypassed altogether.

So, at the expense of adding extra fields one can avoid a nested Conjugate Gradient procedure. This would be particularly important when dynamical fermions are simulated. The chain version of the direct truncation of the Overlap Dirac Operator is similar in appearance to another truncation, usually referred to as "domain walls" [13].

The proposal above, employing chains, has two potential advantages over domain walls. First, it is much more flexible, allowing one to change both the rational approximation one uses and its chain implementation. This flexibility ought to allow various improvements. Second, since here the argument of the approximated sign function is λH_W, not the rather cumbersome logarithm of the transfer matrix of the domain wall case, eigenstates of H_W with small eigenvalues can be eliminated by projection with greater ease. This elimination, although costly numerically, vastly increases the accuracy of the approximation to the sign function. Actually, at this stage of the game and at practical gauge coupling values, the use of projectors seems to be numerically indispensable to direct implementations of the Overlap Dirac Operator. Although no projectors have been implemented in domain wall simulations and physical results have been claimed, to me it seems doubtful that the domain wall version of truncating the Overlap will really be capable to get to small quark masses without employing some technique equivalent to projection.

5   Acknowledgments

I am grateful to the organizers of the Interdisciplinary Workshop on Numerical Challenges to Lattice QCD for the opportunity to participate and to describe the Overlap. I wish to thank them for their warm hospitality. My research at Rutgers is supported by the DOE under grant # DE-FG05-96ER40559.

References

1. R. Narayanan, H. Neuberger, *Phys. Lett.* B **302**, 62 (1993); *Nucl. Phys.* B **412**, 574 (1994); *Nucl. Phys.* B **443**, 305 (1995); *Phys. Lett.* B **393**, 360 (1997); *Nucl. Phys.* B **477**, 521 (1996); *Phys. Lett.* B **348**, 549 (1995); *Phys. Rev. Lett.* **71**, 3251 (1993); Y. Kikuakwa, R. Narayanan, H. Neuberger, *Phys. Rev.* D **57**, 1233 (1998); *Phys. Lett.* B **105**, 399 (1997); S. Randjbar-Daemi, J. Strathdee, *Phys. Lett.* B **348**, 543 (1995); *Nucl. Phys.* B **443**, 386 (1995); *Phys. Lett.* B **402**, 134 (1997); *Nucl. Phys.* B **461**, 305 (1996); *Nucl. Phys.* B **466**, 335 (1996) H. Neuberger, *Phys. Lett.* B **417**, 141 (1998); *Phys. Rev.* D **57**, 5417 (1998); *Phys. Lett.* B **427**, 353 (1998); *Phys. Rev.* D **59**, 085006 (1999); *Phys. Lett.* B **437**, 117 (1998); R. Narayanan, H. Neuberger, P. Vranas, *Phys. Lett.* B **353**, 507 (1995).
2. D. B. Kaplan, *Phys. Lett.* B **288**, 342 (1992); S. A. Frolov, A. A. Slavnov, *Phys. Lett.* B **309**, 344 (1993).
3. H. B. Nielsen, S. E. Ugh, *Nucl. Phys.* B *(Proc. Suppl.)* **29B,C**, 200 (1992) and references therein.
4. P. Ginsparg, K. Wilson, *Phys. Rev.* D **25**, 2649 (1982).
5. N. J. Higham, in Proceedings of "Pure and Applied Linear Algebra: The New Generation", Pensacola, March 1993.

6. K. Jansen, these proceedings.
7. H. Neuberger,Phys. *Phys. Rev. Lett.* **81**, 4060 (1998).
8. A. Frommer, S. Güsken, T. Lippert, B. Nöckel, K. Schilling, *Int. J. Mod. Phys.* C **6**, 627 (1995).
9. H. Neuberger, hep-lat/9811019, *Int. J. Mod. Phys.* C , (), to appear.
10. A. Borici, these proceedings.
11. R. G. Edwards, U. M. Heller, R. Narayanan, NPB **540**, 457 (1999); PRD **59**, 094510 (1999).
12. H. Neuberger, PRD **60**, 065006 (1999).
13. P. Chen et. al., PRD **59**, 054508 (1999).

Solution of $f(A)x = b$ with Projection Methods for the Matrix A

Henk A. van der Vorst

Utrecht Universtity
Department of Mathematics
Utrecht, The Netherlands

Abstract. In this paper, we expand on an idea for using Krylov subspace information for the matrix A and the vector b. This subspace can be used for the approximate solution of a linear system $f(A)x = b$, where f is some analytic function. We will make suggestions on how to use this for the case where f is the matrix *sign* function.

1   Introduction

The matrix *sign* function plays an important role in QCD computations, see for instance [12]. In the computational models one has to compute an approximate solution for linear systems of the type

$$(B + \text{sign}(A))x = b, \tag{1}$$

with $A, B \in \mathbb{C}^{n \times n}$, and A and B do not commute. The latter property is an important bottleneck for the efficient computation of subspaces that can be used for the reduction of both A and B.

In [15] an approach was suggested for the usage of a Krylov subspace for the matrix A and a given vector, for instance b, for the computation of approximate solutions of linear systems

$$f(A)x = b,$$

with f an analytic function. The approach in [15] was motivated by the function $f(A) = A^2$, which plays a role in the solution of some biharmonic systems. Furthermore, the proposed methods were outlined for the case that $A = A^H$. However, the approach is easily generalized for non-symmetric complex matrices, as we will see in this paper. We have to pay more attention to the evaluation of f for the reduced system, associated with the Krylov subspace.

In particular, we will discuss some possible approaches for using the Krylov subspace for the computation of $\text{sign}(A)p$ for given vectors p. With the evaluation of the matrix *sign* function one has to be extremely careful. A popular approach, based on a Newton iteration converges fast, but is sensitive for rounding errors, especially when A is ill-conditioned. We will briefly discuss a computational method that was suggested (and analyzed) by Bai and Demmel [3]. This approach can also be combined, in principle, with the subspace reduction technique. Our early experiments, not reported here, indicate that the actual computation of approximate solutions of (1) is complicated because of the occurrence of the matrix B. Since this matrix does not share an eigenvector basis with A, there is little hope that the subspace generated with A can also be used efficiently with B. It seems that we have to experiment with either nested approaches or with mixed subspaces. The current state of the art is that we still have a way to go in our quest for an efficient computational technique for the very large systems that arise in QCD modeling.

2   Krylov Subspaces

Krylov subspace methods are well-established methods for the reduction of large linear systems of equations to much smaller size problems. We will explain briefly the idea behind Krylov subspace methods. Given a linear system $Ax = b$, with a large, usually sparse, unsymmetric nonsingular matrix A, then the standard Richardson iteration

$$x_k = (I - A)x_{k-1} + b$$

generates approximate solutions in shifted Krylov subspaces

$$x_0 + K^k(A; r_0) = x_0 + \{r_0, Ar_0, \dots, A^{k-1}r_0\},$$

with $r_0 = b - Ax_0$, for some given initial vector x_0.
The Krylov subspace projection methods fall in three different classes:

1. The *Ritz-Galerkin approach*: Construct the x_k for which the residual is orthogonal to the current subspace: $b - Ax_k \perp K^k(A; r_0)$.
2. The *minimum residual approach*: Identify the x_k for which the Euclidean norm $\|b - Ax_k\|_2$ is minimal over $K^k(A; r_0)$.
3. The *Petrov-Galerkin approach*: Find an x_k so that the residual $b - Ax_k$ is orthogonal to some other suitable k-dimensional subspace.

The Ritz-Galerkin approach leads to such popular and well-known methods as Conjugate Gradients, the Lanczos method, FOM, and GENCG. The minimum residual approach leads to methods like GMRES, MINRES, and ORTHODIR. If we select the k-dimensional subspace in the third approach as $K^k(A^H; s_0)$, then we obtain the Bi-CG, and QMR methods. More recently, hybrids of the three approaches have been proposed, like CGS, Bi-CGSTAB, BiCGSTAB(ℓ), TFQMR, FGMRES, and GMRESR.

A nice overview of Krylov subspace methods, with focus on Lanczos-based methods, is given in [7]. Simple algorithms and unsophisticated software for some of these methods is provided in [4]. Iterative methods with much attention to various forms of preconditioning have been described in [2]. A good overview on iterative methods was published by Saad [14]; it is very algorithm oriented, with, of course, a focus on GMRES and preconditioning techniques, like threshold ILU, ILU with pivoting, and incomplete LQ factorizations. An annotated entrance to the vast literature on preconditioned iterative methods is given in [5].

In order to identify optimal approximate solutions in the Krylov subspace we need a suitable basis for this subspace, one that can be extended in a meaningful way for subspaces of increasing dimension. The obvious basis r_0, Ar_0, ..., $A^{i-1}r_0$, for $K^i(A; r_0)$, is not very attractive from a numerical point of view, since the vectors $A^j r_0$ point more and more in the direction of the dominant eigenvector for increasing j and hence the basis vectors will have small mutual angles. This leads to numerically unstable processes.

Instead of the standard basis one usually prefers an orthonormal basis, and Arnoldi [1] suggested to compute this basis as follows. Start with $v_1 \equiv r_0/\|r_0\|_2$. Assume that we have already an orthonormal basis v_1, ..., v_j for $K^j(A; r_0)$, then this basis is expanded by computing $t = Av_j$, and by orthonormalizing this vector t with respect to v_1, ..., v_j. In principle the orthonormalization process can be carried out in different ways, but the most commonly used approach is to do this by a modified Gram-Schmidt procedure [9]. This leads to an algorithm for the creation of an orthonormal basis for $K^m(A; r_0)$, as in Fig 1.

$$v_1 = r_0/\|r_0\|_2;$$
$$\text{for } j = 1, .., m - 1$$
$$\quad t = Av_j;$$
$$\quad \text{for } i = 1, ..., j$$
$$\quad\quad h_{i,j} = v_i^H t;$$
$$\quad\quad t = t - h_{i,j}v_i;$$
$$\quad \text{end};$$
$$\quad h_{j+1,j} = \|t\|_2;$$
$$\quad v_{j+1} = t/h_{j+1,j};$$
$$\text{end}$$

Fig. 1. Arnoldi's method with modified Gram–Schmidt orthogonalization

It is easily verified that v_1, ..., v_m form an orthonormal basis for the Krylov subspace $K^m(A; r_0)$ (that is, if the construction does not terminate at a vector $t = 0$). The orthogonalization leads to relations between the v_j, that can be formulated in a compact algebraic form. Let V_j denote the matrix with columns v_1 up to v_j, then it follows that

$$AV_{m-1} = V_m H_{m,m-1}. \qquad (2)$$

The m by $m-1$ matrix $H_{m,m-1}$ is upper Hessenberg, and its elements $h_{i,j}$ are defined by the Arnoldi algorithm.

From a computational point of view, this construction is composed from three basic elements: a matrix vector product with A, innerproducts, and updates. We see that this orthogonalization becomes increasingly expensive for increasing dimension of the subspace, since the computation of each $h_{i,j}$ requires an inner product and a vector update.

Note that if A is symmetric, then so is $H_{m-1,m-1} = V_{m-1}^H A V_{m-1}$, so that in this situation $H_{m-1,m-1}$ is tridiagonal. This means that in the orthogonalization process, each new vector has to be orthogonalized with respect to the previous two vectors only, since all other innerproducts vanish. The resulting three term recurrence relation for the basis vectors of $K^m(A; r_0)$ is known as the *Lanczos method* and some very elegant methods are derived from it. In the symmetric case the orthogonalization process involves constant arithmetical costs per iteration step: one matrix vector product, two innerproducts, and two vector updates.

3 Reduced Systems

With equation (2) we can construct approximate solutions for $Ax = b$ in the Krylov subspace $K^m(A; r_0)$. These approximate solutions can be written as $x_m = x_0 + V_m y$, with $y \in \mathbb{R}^n$, since the columns of V_m span a basis for the Krylov subspace. The Ritz-Galerkin orthogonality condition for the residual leads to

$$b - A x_m \perp \{v_1, \ldots, v_m\},$$

or

$$V_m^H (b - A(x_0 + V_m y)) = 0.$$

Now we use that $b - A x_0 = r_0 = \|r_0\|_2 v_1$, and with (2) we obtain

$$H_{m,m} y = \|r_0\| e_1, \qquad (3)$$

with e_1 the first canonical basis vector in $\mathbb{R}^m$. If $H_{m,m}$ is not singular then we can write the approximate solution x_m as

$$x_m = \|r_0\|_2 V_m H_{m,m}^{-1} e_1. \qquad (4)$$

Note that this expression closely resembles the expression $x = A^{-1}b$ for the exact solution of $Ax = b$. The matrix $H_{m,m}$ can be interpreted as the restriction of A with respect to $v_1, \ldots, v_m$. The vector $\|r_0\| e_1$ is the expression for the right-hand side with respect to this basis, and V_m is the operator that expresses the solution of the reduced system (in $\mathbb{R}^m$) in terms of the canonical basis for $\mathbb{R}^n$.

Let us from now on assume that $x_0 = 0$. This simplifies the formulas, but is does not pose any further restriction. With $x_0 \neq 0$ we have for $x = w + x_0$ that $A(w + x_0) = b$ or $Aw = b - x_0 = \tilde{b}$, and we are again in the situation that the initial approximation w_0 for w is $w_0 = 0$.

We can also use the above mechanism for the solution of more complicated systems of equations. Suppose that we want to find approximate solutions for $A^2 x = b$, with only the Krylov subspace for A and $r_0 = b$ available. The solution of $A^2 x = b$ can be realized in two steps

1. Solve z_m from $Az = b$, using the Ritz-Galerkin condition. With $z = V_m y$ and (2), we have that

$$z = \|b\|_2 V_m H_{m,m}^{-1} e_1.$$

2. Solve x_m from $Ax_m = z_m$, with $x_m = V_m u$. It follows that

$$AV_m u = \|b\|_2 V_m H_{m,m}^{-1} e_1,$$

$$V_{m+1} H_{m+1,m} u_m = \|b\|_2 V_m h_{m,m}^{-1}.$$

The Ritz-Galerkin condition with respect to V_m leads to

$$H_{m,m} u_m = \|b\|_2 H_{m,m}^{-1} e_1.$$

These two steps lead to the approximate solution

$$x_m = \|b\|_2 V_m H_{m,m}^{-2} e_1. \tag{5}$$

If we compare (4) for $Ax = b$ with (5) for $A^2 x = b$, then we see that the operation with A^2 translates to an operation with $H_{m,m}^2$ for the reduced system and that is all.

Note that this approximate solution x_m does not satisfy a Ritz-Galerkin condition for the system $A^2 x = b$. Indeed, for $x_m = V_m y$, we have that

$$A^2 V_m y = AV_{m+1} H_{m+1,m} y = V_{m+2} H_{m+2,m+1} H_{m+1,m} y.$$

The Ritz-Galerkin condition with respect to V_m, for $b - Ax_m$, leads to

$$V_m^H V_{m+2} H_{m+2,m+1} H_{m+1,m} y = \|b\|_2 e_1.$$

A straight-forward evaluation of $H_{m+2,m+1} H_{m+1,m}$ and the orthogonality of the v_j's, leads to

$$V_m^H V_{m+2} H_{m+2,m+1} H_{m+1,m} = H_{m,m}^2 + h_{m+1,m} h_{m,m+1} e_m e_m^T.$$

That means that the reduced matrix for A^2, expressed with respect to the V_m basis, is given by the matrix $H^2_{m,m}$ in which the bottom right element $h_{m,m}$ is updated with $h_{m+1,m}h_{m,m+1}$. By computing x_m as in (5), we have ignored the factor $h_{m+1,m}h_{m,m+1}$. This is acceptable, since in generic situations the convergence of the Krylov solution process for $Ax = b$ goes hand in hand with small elements $h_{m+1,m}$.

We can go one step further, and try to solve

$$(A^2 + \alpha A + \beta I)x = b,$$

with Krylov subspace information obtained for $Ax = b$ (with $x_0 = 0$). The Krylov subspace $K^m(A, r_0)$ is shift invariant, that is

$$K^m(A, r_0) = K^m(A - \sigma I, r_0),$$

for any scalar $\sigma \in \mathbb{C}$. The matrix polynomial $A^2 + \alpha A + \beta I$ can be factored into

$$A^2 + \alpha A + \beta I = (A - \omega_1 I)(A - \omega_2 I).$$

Proceeding as for A^2, that is solving the given system in two steps, and imposing a Ritz-Galerkin condition for each step, leads to the approximate solution

$$x_m = \|b\|_2 V_m (H^2_{m,m} + \alpha H_{m,m} + \beta I_m)^{-1} e_1.$$

The generalization to higher degree polynomial systems

$$p_n(A)x \equiv (A^n + \alpha_{n-1} A^{n-1} + \ldots + \alpha_0 I)\, x = b$$

is straight forward and leads to an approximate solution of the form

$$x_m = \|b\|_2 V_m p_n(H_{m,m})^{-1} e_1.$$

If f is an analytic function, then we can compute the following approximate solution x_m for the solution of $f(A)x = b$:

$$x_m = \|b\|_2 V_m f(H_{m,m})^{-1} e_1. \tag{6}$$

All these approximations are equal to the exact solution if $h_{m+1,m} = 0$. Because $h_{n+1,n} = 0$, we have the exact solution after at most n iterations. The hope is, of course, that the approximate solutions are sufficiently good after $m \ll n$ iterations. There is little to control the residual for the approximate solutions, since in general $f(A)$ may be an expensive function. We use the Krylov subspace reduction in order to avoid expensive evaluation of $f(A)p$ for $p \in \mathbb{R}^n$. A possibility is to compare successive approximations x_m and to base a stopping criterion on this comparison.

4   Computation of the Inverse of $f(H_{m,m})$

The obvious way of computing $f(H_{m,m})^{-1}$ is to reduce $H_{m,m}$ first to some convenient canonical form, for instance to diagonal form. If $H_{m,m}$ is symmetric (in that case $H_{m,m}$ is tridiagonal) then it can be orthogonally transformed to diagonal form:

$$Q_m^H H_{m,m} Q_m = D_m,$$

with Q_m an m by m orthogonal matrix and D_m a diagonal matrix. We then have that

$$Q^H f(H_{m,m})^{-1} Q = f(D_m)^{-1},$$

and this can be used for an efficient and stable computation of x_m. If $H_{m,m}$ is neither symmetric nor (close to) normal (that is $H_{m,m}^H H_{m,m} = H_{m,m} H_{m,m}^H$, then the transformation to diagonal form cannot be done by an orthogonal operator. If $H_{m,m}$ has no Jordan blocks, the transformation can be done by

$$X_m^{-1} H_{m,m} X_m = D_m.$$

This decomposition is not advisable if the condition number of X_m is much larger than 1. In that case it is much better to reduce the matrix $H_{m,m}$ to Schur form:

$$Q_m^H H_{m,m} Q_m = U_m,$$

with U_m an upper triangular matrix. The eigenvalues of $H_{m,m}$ appear along the diagonal of U_m. If A is real, then the computations can be kept in real arithmetic if we use the property that $H_{m,m}$ can be orthogonally transformed to generalized Schur form. In a generalized Schur form, the matrix U_m may have two by two non-zero blocks along the diagonal (but its strict lower triangular part is otherwise zero). These two by two blocks represent complex conjugate eigenpairs of $H_{m,m}$. For further details on Schur forms, generalized Schur forms, and their computation, see [9].

5   Numerical Examples

Our numerical examples have been taken from [15]. These experiments have been carried out for diagonal real matrices A, which does not mean a loss of generality. In exact arithmetic the Krylov subspace generated with A, and v, coincides with the Krylov subspace generated with $Q^H A Q$ and $Q^H v$, in the sense that

$$K^m(A; v) = Q K^m(Q^H A Q, Q^H v),$$

if Q is an orthogonal matrix ($Q^H Q = I$). In other words, transformation with an orthogonal Q leads to another orientation of the orthogonal basis, but the Krylov subspace method leads to the same approximate solutions. Of course, round-off patterns may be different, but round-off does not have dramatic effects on Krylov subspace methods other than an occasional small delay in the number of iterations. Therefore, we may expect similar behavior for more general matrices with the same eigenvalue distribution.

m	$\|b - A^2 x_m^{new}\|_2$	$\|b - A^2 x_m^{old}\|_2$
0	$0.21E2$	$0.21E2$
10	0.18	0.15
20	$0.27E - 2$	$0.16E - 1$
30	$0.53E - 5$	$0.63E - 2$
40	$0.16E - 8$	$0.28E - 2$
50		$0.36E - 2$
60		$0.10E - 2$
70		$0.49E - 4$
80		$0.18e - 5$
100		$0.21e - 8$

Table 1. Residual norms for approaches A and B

The diagonal matrix A is of order 900. Its eigenvalues are 0.034, 0.082, 0.127, 0.155, 0.190. The remaining 895 eigenvalues are uniformly distributed over the interval $[0.2, 1.2]$. This type of eigenvalue distribution is more or less what one might get with preconditioned Poisson operators. Now suppose that we want to solve $A^2 x = b$, with b a vector with all ones. We list the results for two different approaches:

A We generate the Krylov subspace with A and b and determine the approximate solution x_m^{new} as in (5).

B We generate the Krylov subspace for the operator A^2 and the vector b (the 'classical' approach). This leads to approximations denoted as x_m^{old}.

In Table 1 we have listed the norms of the residuals for the two approaches for some values of m. The analysis in [15] shows that the much faster convergence for the new approach could have been expected. Note that the new approach also has the advantage that there are only m matrix vector products with A for the new approach. For the classical approach we need $2m$ matrix vector products with A, assuming that vectors like $A^2 p$ are computed by applying A twice. Usually, the matrix vector product is the CPU-dominating factor in the computations, since they operate in $\mathbb{R}^n$. The oprations with $H_{m,m}$ are carried out in $\mathbb{R}^m$, and in typical applications $m \ll n$.

In [15] also an example is given for a more complicated function of A, namely the solution of

$$e^A x = b,$$

with A the same diagonal matrix as in the previous example, and b again the vector with all ones. This is a type of problem that one encounters in the solution of linear systems of ODEs. With the Krylov subspace for A and r_0 of dimension 20, a residual

$$||r_m||_2 \equiv ||b - e^A x_m||_2 \approx 8.7E - 12$$

was observed, working in 48 bits floating point precision.

Others have also suggested to work with the reduced system for the computation of, for instance, the exp function of a matrix, as part of solution schemes for (parabolic) systems of equations. See, e.g. [10,8,11].

6   Matrix Sign Function

The matrix sign function $sign(A)$ for a nonsingular matrix A, with no eigenvalues on the imaginary axis, is defined as follows [3,13]. Let

$$A = X \operatorname{diag}(J_+, J_-) X^{-1}$$

denote the decomposition of $A \in \mathbb{C}^{n \times n}$. The eigenvalues of J_+ lie in the right half plane, and those of J_- are in the left half plane. Let I_+ denote the identity matrix with the same dimensions as J_+, and I_- the identity matrix corresponding to J_-. Then

$$\operatorname{sign}(A) \equiv X \operatorname{diag}(I_+, -I_-) X^{-1}.$$

The sign function can be used, amongst others, to compute invariant subspaces, for instance those corresponding to the eigenvalues of A with positive real parts. It plays also an important role in QCD (cf [12]). The Jordan decomposition of A is not a useful vehicle for the computation of this function. It can be shown that $\operatorname{sign}(A)$ is the limit of the Newton iteration

$$A_{k+1} = \frac{1}{2}(A_k + A_k^{-1}), \quad \text{for} \quad k = 0, 1, \ldots, \quad \text{with} \quad A_0 = A.$$

see [3]. Unfortunately, the Newton iteration is also not suitable for accurate computation if A is ill-conditioned. Bai and Demmel consider more accurate ways of computation, which rely on the (block) Schur form of A:

$$B = Q^H A Q = \begin{bmatrix} B_{11} & B_{12} \\ 0 & B_{22} \end{bmatrix}.$$

The matrix Q is orthonormal, and it can easily be shown that

$$\operatorname{sign}(A) = Q \operatorname{sign}(B) Q^H.$$

Let this decomposition be such that B_{11} contains the eigenvalues of A with positive real part. Then let R be the solution of the Sylvester equation

$$B_{11} R - R B_{22} = -B_{12}.$$

This Sylvester equation can also be solved by a Newton iteration process. Then Bai and Demmel proved that

$$\text{sign}(A) = Q \begin{bmatrix} I & -2R \\ 0 & -I \end{bmatrix} Q^H.$$

See [3] for further details, stability analysis, and examples of actual computations.

Thomas Lippert [1] has experimented with alternative schemes for the computation of $\text{sign}(A)$, in particular the Schultz iteration

$$A_{k+1} = \frac{1}{2} A_k (3I - A_k^2).$$

We do not know of any stability analysis for this approach; our own preliminary experiments suggest a rather slow convergence.

Suppose that we want to solve $\text{sign}(A)x = b$. Then, in view of the previous section, we suggest to start with constructing the Krylov subspace $K^m(A; b)$, and to compute the sign function for the reduced matrix $H_{m,m}$. This leads to the following approximation for the solution of $\text{sign}(A) = b$:

$$x_m = \|b\|_2 V_m \text{sign}(H_{m,m})^{-1} e_1. \tag{7}$$

Our preliminary experiments with this approach are encouraging. The actual problem in QCD, however, is often to solve an essentially different equation

$$(C + \text{sign}(A))x = b.$$

See, for instance, [12]. We have not yet succeeded in identifying efficient computational schemes for this shifted equation. An obvious way to solve this equation is to apply a nested Krylov solution process. In the outer process one forms the Krylov subspace for $C + \text{sign}(A)$ and b, and for each matrix vector product involved in this process one computes approximations for $\text{sign}(A)p$, for given vector v, using the process that we have outlined above. Our experiments, not reported here, show that this is a very expensive process. We are currently experimenting with nested Newton iterations for the inverse of $C + \text{sign}(A)$ and the sign function. The idea is to carry out these Newton processes in an inexact way and to accelerate the process, in a way that has been described in [6].

References

1. Arnoldi, W. E.: The principle of minimized iteration in the solution of the matrix eigenproblem. *Quart. Appl. Math.* **9** (1951) 17–29
2. Axelsson, O.: *Iterative Solution Methods.* Cambridge University Press, Cambridge (1994)

---

[1] Private communication

3. Bai, Z., Demmel, J.: Using the matrix sign function to compute invariant subspaces. *SIAM J. Matrix Anal. Applic.* **19** (1998) 205–225
4. Barrett, R., Berry, M., Chan, T., Demmel, J., Donato, J., Dongarra, J., Eijkhout, V., Pozo, R., Romine, C., van der Vorst, H.: *Templates for the Solution of Linear Systems: Building Blocks for Iterative Methods.* SIAM, Philadelphia, PA (1994)
5. Bruaset, A.M.: *A Survey of Preconditioned Iterative Methods.* Longman Scientific & and Technical, Harlow, UK (1995)
6. Fokkema, D.R., Sleijpen, G.L.G., van der Vorst, H.A.: Accelerated inexact Newton schemes for large systems of nonlinear equations. *SIAM J. Sci. Comput.* **19** (1998) 657–674
7. Freund, R.W., Golub, G.H., Nachtigal, N.M.: Iterative solution of linear systems. In *Acta Numerica 1992.* Cambridge University Press, Cambridge (1992)
8. Gallopoulos, E., Saad, Y.: Efficient solution of parabolic equations by Krylov approximation methods. *SIAM J. Sci. Statist. Comput.* **13** (1992) 1236–1264
9. Golub, G.H., Van Loan, C.F.: *Matrix Computations.* The Johns Hopkins University Press, Baltimore (1996)
10. Hochbruck, M., Lubich, C.: On Krylov subspace approximations to the matrix exponential operator. *SIAM J. Numer. Anal.* **34** (1997) 1911–1925
11. Meerbergen, K., Sadkane, M.: Using Krylov approximations to the matrix exponential operator in Davidson's method. *Appl. Numer. Math.* **31** (1999) 331–351
12. Neuberger, H.: The Overlap Dirac Operator. in: Frommer, A., Lippert, Th., Medeke, B., Schilling, K. (edts.). Numerical Challenges in Lattice Quantum Chromodynamics. Proceedings of the Interdisciplinary Workshop on Numerical Challenges in Lattice QCD, Wuppertal, August 22-24, 1999. Series Lecture Notes in Computational Science and Engineering (LNCSE). Springer Verlag, Heidelberg (2000)
13. Roberts, J.: Linear model reduction and solution of the algebraic Riccati equation. *Inter. J. Control* **32** (1980) 677–687
14. Saad, Y.: *Iterative Methods for Sparse Linear Systems.* PWS Publishing Company, Boston (1996)
15. van der Vorst, H.A.: An iterative solution method for solving $f(A)x = b$, using Krylov subspace information obtained for the symmetric positive definite matrix A. *J. Comp. and Appl. Math.* **18** (1987) 249–263

A Numerical Treatment of Neuberger's Lattice Dirac Operator

Pilar Hernández[1], Karl Jansen[1†], and Laurent Lellouch[2‡]

[1] CERN
1211 Geneva 23
Switzerland
[2] LAPTH
Chemin de Bellevue, B.P. 110
74941 Annecy-le-Vieux Cedex, France

Abstract. We describe in some detail our numerical treatment of Neuberger's lattice Dirac operator as implemented in a practical application. We discuss the improvements we have found to accelerate the numerical computations and give an estimate of the expense when using this operator in practice.

1 Lattice Formulation of QCD

Today, we believe that the world of quarks and gluons is described theoretically by quantum chromodynamics (QCD). This model shows a number of non-perturbative aspects that cannot be adequately addressed by approximation schemes such as perturbation theory. The only way to evaluate QCD, addressing both its perturbative and non-perturbative aspects *at the same time*, is lattice QCD. In this approach the theory is put on a 4-dimensional Euclidean space-time lattice of finite physical length L, with a non-vanishing value of the lattice spacing a. Having only a finite number of grid points, physical quantities can be computed numerically by solving a high-dimensional integral by Monte Carlo methods, making use of importance sampling.

The introduction of a lattice spacing regularizes the theory and is an intermediate step in the computation of physical observables. Eventually, the regularization has to be removed and the value of the lattice spacing has to be sent to zero to reach the target theory, i.e. continuum QCD. In fact, in conventional formulations of lattice QCD [1], the introduction of the lattice spacing renders the theory on the lattice somewhat different from the continuum analogue and a number of properties of the continuum theory are only very difficult and cumbersome to establish in the lattice regularized

---

† Heisenberg foundation fellow.
‡ On leave from: Centre de Physique Théorique, Case 907, CNRS Luminy, F-13288 Marseille Cedex 9, France.

theory. One of the main reasons for this difficulty is that in conventional lattice QCD the regularization breaks a particular symmetry of the continuum theory, which plays a most important role there, namely chiral symmetry.

However, the last few years have seen a major breakthrough in that we now have formulations of lattice QCD that have an exact lattice chiral symmetry [3]. In this approach, many properties of continuum QCD are preserved *even at a non-vanishing value of the lattice spacing* [2,4,5,3,6]. This development followed the rediscovery [7] of the so-called Ginsparg–Wilson (GW) relation [8] which is fulfilled by any operator with the exact lattice chiral symmetry of [3]. It is not the aim of this contribution to discuss the physics consequences of the GW relation. We have to refer the interested reader to reviews [9,10] about these topics. Here we would like to discuss the *numerical* treatment of a particular lattice operator that satisfies the GW relation, namely Neuberger's solution [5]. This solution has a complicated structure and is challenging to implement numerically. Thus, the large theoretical advantage of an operator satisfying the GW relation must be weighed against the very demanding computational effort required to implement it.

This contribution is organized as follows. After discussing Neuberger's lattice Dirac operator we want to show how we evaluated the operator in our practical application [11] and what kind of improvements we found to accelerate the numerical computations. For alternative ideas for improvements, see the contributions of H. Neuberger [12] and A. Borici [13] to this workshop. We finally give some estimates of the computational expense of using Neuberger's operator.

2   Neuberger's Lattice Dirac Operator

The operator we have used acts on fields (complex vectors) $\Phi(x)$ where $x = (x_0, x_1, x_2, x_3)$ and the $x_\mu, \mu = 0, 1, 2, 3$, are integer numbers denoting a 4-dimensional grid point in a lattice of size N^4 with $N = L/a$. The fields $\Phi(x)$ carry in addition a "colour" index $\alpha = 1, 2, 3$ as well as a "Dirac" index $i = 1, 2, 3, 4$. Hence, Φ is a $N^4 \cdot 3 \cdot 4$ complex vector.

In order to reach the expression for Neuberger's operator we first introduce the matrix A

$$A = 1 + s - \frac{a}{2} \left\{ \gamma_\mu \left(\nabla_\mu^* + \nabla_\mu \right) - a \nabla_\mu^* \nabla_\mu \right\} , \tag{1}$$

where ∇_μ and ∇_μ^* are the nearest-neighbour forward and backward derivatives, the precise definition of which can be found in the Appendix. The parameter s is to be taken in the range of $|s| < 1$ and serves to optimize the localization properties [14] of Neuberger's operator, which is then given by

$$D = \frac{1}{a} \left\{ 1 - A \left(A^\dagger A \right)^{-1/2} \right\} . \tag{2}$$

Through the appearance of the square root in eq. (2), all points on the lattice are connected with one another, giving rise to a very complicated, multi-neighbour action. However, the application of D to a vector Φ will only contain applications of A or $A^\dagger A$ on this vector. Since these matrices are sparse, as only nearest-neighbour interactions are involved, we will never have to store the whole matrix.

In the computation of physical quantities, the inverse of D, applied to a given vector, is generically needed. Hence one faces the problem of having to compute a vector $X = D^{-1}\eta$, with η a prescribed vector (the "source") as required by the particular problem under investigation. Fortunately, a number of efficiently working algorithms for computing $X = D^{-1}\eta$ are known, such as conjugate gradient, BiCGstab, or variants thereof [16]. In conventional approaches to lattice QCD an operator $\tilde{D}$ is used that is very similar to the matrix A in eq. (1). Computing the vector $\tilde{X} = \tilde{D}^{-1}\eta$ requires a number n_{iter} of iterations of some particular method, say BiCGstab. Employing Neuberger's operator D in computing $X = D^{-1}\eta$, it turns out that the number of iterations needed is of the same order of magnitude as when using $\tilde{D}$. At the same time, in each of these iterations, the square root has to be evaluated. When this is done by some polynomial approximation, it is found that the required degree of this polynomial is roughly of the same order as the number of iterations needed for computing the vector X. Hence, with respect to the conventional case, the numerical effort is squared and the price to pay for using the operator D is high.

On the other hand, any solution of the Ginsparg–Wilson relation gives us a tool by which particular problems in lattice QCD can be studied, which would be extremely hard to address with conventional approaches. It is for these cases that the large numerical effort is justified, but clearly, we would like to have clever ideas coming from areas such as Applied Mathematics, to decrease the numerical expense or even overcome this bottleneck.

3   Approximation of $\left(A^\dagger A\right)^{-1/2}$

For computing the square root that appears in eq. (2), we have chosen a Chebyshev approximation [15] by constructing a polynomial $P_{n,\epsilon}(x)$ of degree n, which has an exponential convergence rate in the interval $x \in [\epsilon, 1]$. Outside this interval, convergence is still found but it will not be exponential. The advantages of using this Chebyshev approximation are the well-controlled exponential fit accuracy as well as the possibility of having numerically very stable recursion relations [17] to construct the polynomial, allowing for large degrees. In order to have an optimal approximation, it is desirable to know the lowest and the highest eigenvalue of $A^\dagger A$. A typical example of the eigenvalues of $A^\dagger A$ is shown in fig. 1, where we show the 11 lowest eigenvalues as obtained on a number of configurations using the Ritz functional method [19]. There is a wide spread and very low-lying eigenvalues appear. Choosing ϵ to be the

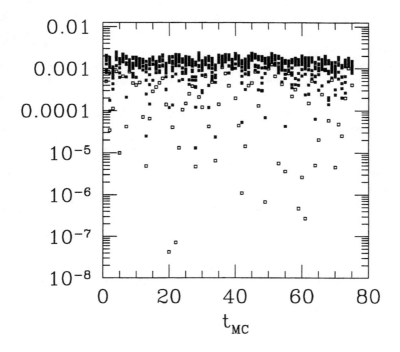

Fig. 1. Monte Carlo time evolution of the eleven lowest eigenvalues of $A^\dagger A$ at $\beta = 5.85$. The lowest eigenvalue for each configuration is the open square.

value of the lowest of these eigenvalues would result in a huge degree n of the polynomial $P_{n,\epsilon}$. We therefore computed O(10) lowest-lying eigenvalues of $A^\dagger A$ as well as their eigenfunctions and projected them out of the matrix $A^\dagger A$. The approximation is then only performed for the matrix with a reduced condition number, resulting in a substantial decrease of the degree of the polynomial. In addition, we computed the highest eigenvalue of $A^\dagger A$ and normalized the matrix A such that $\|A^\dagger A\| \lesssim 1$.

Since our work [11], aiming at the physical question of spontaneous chiral symmetry breaking in lattice QCD, has been one of the first of its kind, we wanted to exclude possible systematic errors and demanded a very high precision for the approximation to the square root:

$$\|X - P_{n,\epsilon}(A^\dagger A)A^\dagger A P_{n,\epsilon}(A^\dagger A)X\|^2/\|2X\|^2 < 10^{-16} \qquad (3)$$

where X is a gaussian random vector. In our practical applications we fixed this precision beforehand and set ϵ to be the 11th lowest eigenvalue of $A^\dagger A$. This then determines the degree of the polynomial n and hence our approximation D_n to the exact Neuberger operator D. We checked that the precision we required for the approximation of the square root is directly related to the precision by which the GW relation itself is fulfilled. Choosing n such that the accuracy in eq. (3) is reached results in

$$\| [\gamma_5 D_n + D_n \gamma_5 - D_n \gamma_5 D_n] X \|^2 / \|X\|^2 \approx 10^{-16} \ . \tag{4}$$

In addition, we find that the deviations from the exact GW relation decrease exponentially fast with increasing n.

4   The Inverse of Neuberger's Operator

As mentioned above, in physics applications a vector $D^{-1}\eta$ has to be computed, with η a prescribed source vector. Not only is the computation of this vector very costly, there also appears to be a conceptual problem: in inspecting the lowest eigenvalue of $D_n^\dagger D_n$, very small eigenvalues are often found as shown in fig. 2. These very small eigenvalues belong to a given chiral sector of the theory, i.e. their corresponding eigenfunctions χ are eigenfunctions of γ_5 with $\gamma_5 \chi = \pm \chi$. In fact, these modes play an important physical role as they are associated with topological sectors of the theory [4,2,3,5].

As far as the practical applications are concerned, it is clear that in the presence of such a small eigenvalue, the inversion of D_n will be very costly, as the condition number of the problem is then very high. In order to address this problem, we followed two strategies:

(i)   We compute the lowest eigenvalue of $D_n^\dagger D_n$ and its eigenfunction (using again the Ritz functional method [19]) and if it is a zero mode—in which case it is also a zero mode of D_n—we project this mode out of D_n and invert only the reduced matrix; this is then well conditioned, as the very small eigenvalues appear to be isolated. In this strategy, the knowledge of the eigenfunction must be very precise and an accuracy of approximating the square root as indicated in eq. (3) is mandatory.

(ii) Again we determine the lowest eigenvalue of $D_n^\dagger D_n$ and the chirality of the corresponding zero mode, if there is any. We then make use of the fact that $D_n^\dagger D_n$ commutes with γ_5. This allows us to perform the inversion in the chiral sector *without zero modes*. In this strategy, the accuracy demanded in eq. (3) could be relaxed and this strategy, which essentially follows ref. [18], is in general much less expensive than following strategy (i).

However, even adopting strategy (ii), solving the system $D_n X = \eta$ is still costly. We therefore tried two ways of improving on this. We first note that

34   Pilar Hernández et al.

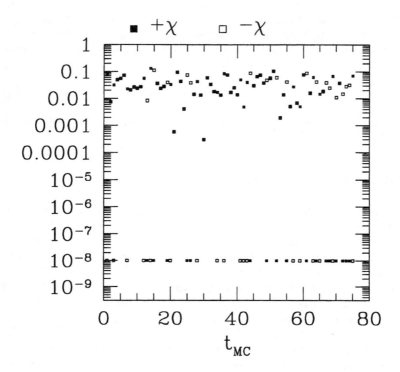

Fig. 2. Monte Carlo time evolution of the lowest eigenvalue of $D_n^\dagger D_n$. The eigenvalues belong to given chiral sectors of the theory denoted as $\pm\chi$ for chirality plus (full squares) and minus (open squares). Data are obtained at $\beta = 5.85$ choosing $s = 0.6$. Whenever there is a zero mode of $D_n^\dagger D_n$, the value of the lowest eigenvalue is set to 10^{-8}.

instead of solving

$$\left[1 - A/\sqrt{A^\dagger A}\right] X = \eta \tag{5}$$

we can equally well solve

$$\left[A^\dagger - \sqrt{A^\dagger A}\right] X = A^\dagger \eta . \tag{6}$$

In practice, however, we found no real advantage in using the formulation of eq. (6). We have further considered two acceleration schemes.

Scheme (a)

We choose two different polynomials (now approximating $\sqrt{A^\dagger A}$ and not the inverse) $P_{n,\epsilon}$ and $P_{m,\epsilon}$, $m < n$, such that

$$P_{n,\epsilon} = P_{m,\epsilon} + \Delta \tag{7}$$

with Δ a "small" correction. Then we have

$$\left[A^\dagger - P_{n,\epsilon}\right]^{-1} = \left[A^\dagger - P_{m,\epsilon} - \Delta\right]^{-1}$$
$$\approx \left[1 + \left(A^\dagger - P_{m,\epsilon}\right)^{-1}\Delta\right]\left(A^\dagger - P_{m,\epsilon}\right)^{-1} . \tag{8}$$

This leads us to the following procedure of solving $D_n X = \eta$:

(1)   first solve

$$\left(A^\dagger - P_{m,\epsilon}\right) Y = \eta ; \tag{9}$$

(2)   then solve

$$\left(A^\dagger - P_{m,\epsilon}\right) X_0 = \eta + \Delta Y ; \tag{10}$$

(3)   use X_0 as a starting vector to finally solve

$$\left(A^\dagger - P_{n,\epsilon}\right) X = \eta . \tag{11}$$

The generation of the starting vector X_0 in steps (1) and (2) is only a small overhead. In fig. 3 we plot the relative residuum $\epsilon_{\text{stop}}^2 = \|D_n X - \eta\|^2 / \|X\|^2$ as a function of the number of applications of D_n. In this case $n = 100$ and $m = 30$. We show the number of applications of the matrix D_n for the case of a random starting vector (dotted line) and the case where X_0 was generated according to the above procedure (solid line). The gain is of approximately a factor of two.

Scheme (b)

In the second approach, we use a sequence of polynomials to solve $D_n X = \eta$. To this end we first solve

$$\left[A^\dagger - P_{m_1,\epsilon}\right] X_1 = A^\dagger \eta \tag{12}$$

by choosing a polynomial $P_{m_1,\epsilon}$ and a stopping criterion for the solver $\epsilon_{\text{stop}}^{(1)}$ such that

$$m_1 < n, \; \epsilon_{\text{stop}}^{(1)} > \epsilon_{\text{stop}} . \tag{13}$$

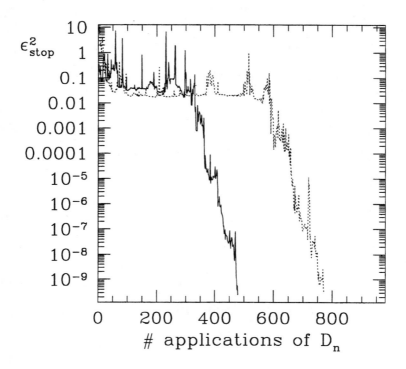

Fig. 3. The residuum as a function of the number of applications of the matrix D_n. The dotted line corresponds to a random starting vector. The solid line to a starting vector generated following scheme (a).

The value of $\epsilon_{\text{stop}}^{(1)}$ is chosen such that it is roughly of the same order of magnitude as the error that the polynomial of degree m_1 itself induces. The solution X_1 is then used as a starting vector for the next equation, employing a polynomial $P_{m_2,\epsilon}$ and stopping criterion $\epsilon_{\text{stop}}^{(2)}$ with

$$m_1 < m_2 < n, \ \epsilon_{\text{stop}}^{(1)} > \epsilon_{\text{stop}}^{(2)} > \epsilon_{\text{stop}} \ . \tag{14}$$

This procedure is then repeated until we reach the desired polynomial $P_{n,\epsilon}$ and stopping criterion ϵ_{stop} to solve the real equation

$$\left[A^\dagger - P_{n,\epsilon}\right] X = A^\dagger \eta \ . \tag{15}$$

As for scheme (a), we gain a factor of about two in the numerical effort. We finally remark that some first tests using the scheme proposed in [13] resulted in a similar performance gain as the two schemes presented above.

In table 1 we give a typical example of the expense of a simulation following strategy (ii). We list both the cost of computing the lowest eigenvalue of $D_n^\dagger D_n$ in terms of the number of iterations to minimize the Ritz functional [19] and the number of iterations to solve $D_n X = \eta$. In both applications, a polynomial of degree n is used to approximate the square root. The numbers in table 1 indicate that a quenched calculation, employing Neuberger's operator, leads to a computational cost that is comparable with a dynamical simulation using conventional operators.

N	n	n_{ev}	n_{invert}
8	190	170	80
10	250	325	200
12	325	700	300

Table 1. N is the number of lattice sites along a side of the hypercube; n, the degree of polynomial; n_{ev}, the number of iterations required to obtain the lowest eigenvalue of $D_n^\dagger D_n$; and n_{invert}, the number of iterations necessary to compute $X = D_n^{-1}\eta$.

5   Conclusions

The theoretical advance that an exact chiral symmetry brings to lattice gauge theory is accompanied by the substantial increase in numerical effort that is required to implement operators satisfying the GW relation. Thus, while the Nielsen–Ninomiya theorem has been circumvented, the "no free lunch theorem" has not. Whether alternative formulations, such as domain wall fermions, can help in this respect remains to be seen.

Appendix

We give here the explicit definitions needed in eq. (1).
The forward and backward derivatives ∇_μ, ∇_μ^* act on a vector $\Phi(x)$ as

$$\nabla_\mu \Phi(x) = \frac{1}{a}\left[U(x,\mu)\Phi(x + a\hat{\mu}) - \Phi(x)\right]$$

$$\nabla_\mu^* \Phi(x) = \frac{1}{a}\left[\Phi(x) - U(x - a\hat{\mu}, \mu)^{-1}\Phi(x - a\hat{\mu})\right] ,$$

where $\hat{\mu}$ denotes the unit vector in direction μ. The (gauge) field $U(x,\mu) \in SU(3)$ lives on the links connecting lattice points x and $x + a\hat{\mu}$ and acts on the colour index $\alpha = 1,2,3$ of the field Φ. Finally, the Dirac matrices γ_μ, $\mu = 0,1,2,3$ are hermitean 4×4 matrices acting on the Dirac index i of the field Φ. Their explicit form is given by

$$\gamma_\mu = \begin{pmatrix} 0 & e_\mu \\ e_\mu^\dagger & 0 \end{pmatrix} \tag{16}$$

with

$$e_0 = 1, \quad e_k = -i\sigma_k \tag{17}$$

and

$$\sigma_1 = \begin{pmatrix} 0 & 1 \\ 1 & 0 \end{pmatrix} \quad \sigma_2 = \begin{pmatrix} 0 & -i \\ i & 0 \end{pmatrix} \quad \sigma_3 = \begin{pmatrix} 1 & 0 \\ 0 & -1 \end{pmatrix} . \tag{18}$$

With the choice of the γ matrices given above, the matrix $\gamma_5 = \gamma_0\gamma_1\gamma_2\gamma_3$ is diagonal and given by

$$\gamma_5 = \begin{pmatrix} 1 & 0 \\ 0 & -1 \end{pmatrix} . \tag{19}$$

We finally note that whenever repeated indices appear, they are summed over.

References

1. K.G. Wilson, Phys. Rev. **D10** (1974) 2445.
2. P. Hasenfratz, V. Laliena and F. Niedermayer, Phys.Lett.**B427** (1998) 125.
3. M. Lüscher, Phys. Lett. **B428** (1998) 342.
4. P. Hasenfratz, Nucl. Phys. **B525** (1998) 401.
5. H. Neuberger, Phys. Lett. **B417** (1998) 141 and **B427** (1998) 353.
6. S. Chandrasekharan, Phys. Rev. **D60** (1999) 074503.
7. P. Hasenfratz, Nucl. Phys. B (Proc.Suppl.) **63A-C** (1998) 53.
8. P.H. Ginsparg and K.G. Wilson, Phys. Rev. **D25** (1982) 2649.
9. F. Niedermayer, Nucl. Phys. B (Proc.Suppl.) **73** (1999) 105.
10. T. Blum, Nucl.Phys.B (Proc.Suppl.) **73** (1999) 167.
11. P. Hernández, K. Jansen and L. Lellouch, CERN preprints CERN-TH/99-197, hep-lat/9907022 and CERN-TH/99-273, hep-lat/9909026.
12. H. Neuberger, hep-lat/9910040.
13. A. Borici, hep-lat/9910045.
14. P. Hernández, K. Jansen and M. Lüscher, Nucl. Phys. **B552** (1999) 363.
15. W. H. Press, S. A. Teukolsky, W. T. Vetterling and B. P. Flannery, *Numerical Recipes*, Second Edition, Cambridge University Press, Cambridge, 1992.
16. Y. Saad, *Iterative methods for sparse linear systems*, PWS Publishing Company, Boston, 1996.

17. L. Fox and I. B. Parker, *Chebyshev polynomials in numerical analysis*, Oxford University Press, London, 1968.
18. R.G. Edwards, U.M. Heller and R. Narayanan, Phys. Rev. **D59** (1999) 094510.
19. B. Bunk, K. Jansen, M. Lüscher and H. Simma, DESY report (September 1994); T. Kalkreuter and H. Simma, Comp.Phys.Comm. **93** (1996) 33.

Fast Methods for Computing the Neuberger Operator

Artan Boriçi

Paul Scherrer Institute
PSI
CH-5232 Villigen

Abstract. I describe a Lanczos method to compute the Neuberger Operator and a multigrid algorithm for its inversion.

1 Introduction

Quantum Chromodynamics (QCD) is a theory of strong interactions, where the chiral symmetry plays a mayor role. There are different starting points to formulate a lattice theory with exact chiral symmetry, but all of them must obey the Ginsparg-Wilson condition [1]:

$$\gamma_5 D^{-1} + D^{-1}\gamma_5 = a\gamma_5\alpha^{-1}, \tag{1}$$

where a is the lattice spacing, D is the lattice Dirac operator and α^{-1} is a local operator and trivial in the Dirac space.

A candidate is the overlap operator of Neuberger [2]:

$$D = 1 - A(A^\dagger A)^{-1/2}, \quad A = M - aD_W \tag{2}$$

where M is a shift parameter in the range $(0,2)$, which I have fixed at one and D_W is the Wilson-Dirac operator,

$$D_W = \frac{1}{2}\sum_\mu [\gamma_\mu(\partial_\mu^* + \partial_\mu) - a\partial_\mu^*\partial_\mu] \tag{3}$$

and ∂_μ and ∂_μ^* are the nearest-neighbor forward and backward difference operators, which are covariant, i.e. the shift operators pick up a unitary 3 by 3 matrix with determinant one. These small matrices are associated with the links of the lattice and are oriented positively. A set of such matrices forms a "configuration". $\gamma_\mu, \mu = 1, \dots, 5$ are 4 by 4 matrices related to the spin of the particle. Therefore, if there are N lattice points, the matrix is of order $12N$. A restive symmetry of the matrix A that comes from the continuum is the so called $\gamma_5 - symmetry$ which is the Hermiticity of the $\gamma_5 A$ operator.

The computation of the inverse square root of a matrix is reviewed in [3]. In the context of lattice QCD there are several sparse matrix methods, which are developed recently [4–8]. I will focus here on a Lanczos method similar to [4]. For a more general case of functions of matrices I refer to the talk of H. van der Vorst, and for a Chebyshev method I refer to the talk of K. Jansen, both included in these proceedings.

2 The Lanczos Algorithm

The Lanczos iteration is known to approximate the spectrum of the underlying matrix in an optimal way and, in particular, it can be used to solve linear systems [9].

Let $Q_n = [q_1, \ldots, q_n]$ be the set of orthonormal vectors, such that

$$A^\dagger A Q_n = Q_n T_n + \beta_n q_{n+1} (e_n^{(n)})^T, \quad q_1 = \rho_1 b, \quad \rho_1 = 1/\|b\|_2 \qquad (4)$$

where T_n is a tridiagonal and symmetric matrix, b is an arbitrary vector, and β_n a real and positive constant. $e_m^{(n)}$ denotes the unit vector with n elements in the direction m.

By writing down the above decomposition in terms of the vectors $q_i, i = 1, \ldots, n$ and the matrix elements of T_n, I arrive at a three term recurrence that allows to compute these vectors in increasing order, starting from the vector q_1. This is the *Lanczos Algorithm*:

$$
\begin{aligned}
&\beta_0 = 0, \ \rho_1 = 1/\|b\|_2, \ q_0 = o, \ q_1 = \rho_1 b \\
&for \ i = 1, \ldots \\
&\quad v = A^\dagger A q_i \\
&\quad \alpha_i = q_i^\dagger v \\
&\quad v := v - q_i \alpha_i - q_{i-1} \beta_{i-1} \\
&\quad \beta_i = \|v\|_2 \\
&\quad if \beta_i < tol, \ n = i, \ end \ for \\
&\quad q_{i+1} = v/\beta_i
\end{aligned}
\qquad (5)
$$

where *tol* is a tolerance which serves as a stopping condition.

The Lanczos Algorithm constructs a basis for the Krylov subspace [9]:

$$\text{span}\{b, A^\dagger A b, \ldots, (A^\dagger A)^{n-1} b\} \qquad (6)$$

If the Algorithm stops after n steps, one says that the associated Krylov subspace is invariant.

In the floating point arithmetic, there is a danger that once the Lanczos Algorithm (polynomial) has approximated well some part of the spectrum, the iteration reproduces vectors which are rich in that direction [9]. As a consequence, the orthogonality of the Lanczos vectors is spoiled with an immediate impact on the history of the iteration: if the algorithm would stop after n steps in exact arithmetic, in the presence of round off errors the loss of orthogonality would keep the algorithm going on.

3   The Lanczos Algorithm for Solving $A^\dagger A x = b$

Here I will use this algorithm to solve linear systems, where the loss of orthogonality will not play a role in the sense that I will use a different stopping condition.

I ask the solution in the form

$$x = Q_n y_n \tag{7}$$

By projecting the original system onto the Krylov subspace I get:

$$Q_n^\dagger A^\dagger A x = Q_n^\dagger b \tag{8}$$

By construction, I have

$$b = Q_n e_1^{(n)}/\rho_1, \tag{9}$$

Substituting $x = Q_n y_n$ and using (4), my task is now to solve the system

$$T_n y_n = e_1^{(n)}/\rho_1 \tag{10}$$

Therefore the solution is given by

$$x = Q_n T_n^{-1} e_1^{(n)}/\rho_1 \tag{11}$$

This way using the Lanczos iteration one reduces the size of the matrix to be inverted. Moreover, since T_n is tridiagonal, one can compute y_n by short recurences.

If I define:

$$r_i = b - A^\dagger A x_i, \quad q_i = \rho_i r_i, \quad y_i = \rho_i x_i \tag{12}$$

where $i = 1, \ldots$, it is easy to show that

$$\begin{aligned} \rho_{i+1}\beta_i + \rho_i\alpha_i + \rho_{i-1}\beta_{i-1} = 0 \\ q_i + y_{i+1}\beta_i + y_i\alpha_i + y_{i-1}\beta_{i-1} = 0 \end{aligned} \tag{13}$$

Therefore the solution can be updated recursively and I have the following *Algorithm1 for solving the system* $A^\dagger A x = b$:

$$\begin{aligned}
&\beta_0 = 0, \ \rho_1 = 1/||b||_2, \ q_0 = o, \ q_1 = \rho_1 b \\
&for \ i = 1, \ldots \\
&\quad v = A^\dagger A q_i \\
&\quad \alpha_i = q_i^\dagger v \\
&\quad v := v - q_i\alpha_i - q_{i-1}\beta_{i-1} \\
&\quad \beta_i = ||v||_2 \\
&\quad q_{i+1} = v/\beta_i \\
&\quad y_{i+1} = -\frac{q_i + y_i\alpha_i + y_{i-1}\beta_{i-1}}{\beta_i} \\
&\quad \rho_{i+1} = -\frac{\rho_i\alpha_i + \rho_{i-1}\beta_{i-1}}{\beta_i} \\
&\quad r_{i+1} := q_{i+1}/\rho_{i+1} \\
&\quad x_{i+1} := y_{i+1}/\rho_{i+1} \\
&\quad if \ \frac{1}{|\rho_{i+1}|} < tol, \ n = i, \ end \ for
\end{aligned} \tag{14}$$

4   The Lanczos Algorithm for Solving $(A^\dagger A)^{1/2}x = b$

Now I would like to compute $x = (A^\dagger A)^{-1/2}b$ and still use the Lanczos Algorithm. In order to do so I make the following observations:

Let $(A^\dagger A)^{-1/2}$ be expressed by a matrix-valued function, for example the integral formula [3]:

$$(A^\dagger A)^{-1/2} = \frac{2}{\pi} \int_0^\infty dt(t^2 + A^\dagger A)^{-1} \tag{15}$$

From the previous section, I use the Lanczos Algorithm to compute

$$(A^\dagger A)^{-1}b = Q_n T_n^{-1} e_1^{(n)}/\rho_1 \tag{16}$$

It is easy to show that the Lanczos Algorithm is shift-invariant. i.e. if the matrix $A^\dagger A$ is shifted by a constant say, t^2, the Lanczos vectors remain invariant. Moreover, the corresponding Lanczos matrix is shifted by the same amount.

This property allows one to solve the system $(t^2 + A^\dagger A)x = b$ by using the same Lanczos iteration as before. Since the matrix $(t^2 + A^\dagger A)$ is better conditioned than $A^\dagger A$, it can be concluded that once the original system is solved, the shifted one is solved too. Therefore I have:

$$(t^2 + A^\dagger A)^{-1}b = Q_n(t^2 + T_n)^{-1}e_1^{(n)}/\rho_1 \tag{17}$$

Using the above integral formula and puting everything together, I get:

$$x = (A^\dagger A)^{-1/2}b = Q_n T_n^{-1/2} e_1^{(n)}/\rho_1 \tag{18}$$

There are some remarks to be made here:

a) By applying the Lanczos iteration on $A^\dagger A$, the problem of computing $(A^\dagger A)^{-1/2}b$ reduces to the problem of computing $y_n = T_n^{-1/2}e_1^{(n)}/\rho_1$ which is typically a much smaller problem than the original one. But since $T_n^{1/2}$ is full, y_n cannot be computed by short recurences. It can be computed for example by using the full decomposition of T_n in its eigenvalues and eigenvectors; in fact this is the method I have employed too, for its compactness and the small overhead for moderate n.

b) The method is not optimal, as it would have been, if one would have applied it directly for the matrix $(A^\dagger A)^{1/2}$. By using $A^\dagger A$ the condition is squared, and one looses a factor of two compared to the theoretical case!

c) From the derivation above, it can be concluded that the system $(A^\dagger A)^{1/2}$ $x = b$ is solved at the same time as the system $A^\dagger A x = b$.

d) To implement the result (18), I first construct the Lanczos matrix and then compute

$$y_n = T_n^{-1/2} e_1^{(n)}/\rho_1 \tag{19}$$

To compute $x = Q_n y_n$, I repeat the Lanczos iteration. I save the scalar products, though it is not necessary.

Therefore I have the following *Algorithm2 for solving the system* $(A^\dagger A)^{1/2}$ $x = b$:

$\beta_0 = 0, \ \rho_1 = 1/\|b\|_2, \ q_0 = o, \ q_1 = \rho_1 b$
for $i = 1, \ldots$
    $v = A^\dagger A q_i$
    $\alpha_i = q_i^\dagger v$
    $v := v - q_i \alpha_i - q_{i-1} \beta_{i-1}$
    $\beta_i = \|v\|_2$
    $q_{i+1} = v/\beta_i$
    $\rho_{i+1} = -\dfrac{\rho_i \alpha_i + \rho_{i-1}\beta_{i-1}}{\beta_i}$
    if $\frac{1}{|\rho_{i+1}|} < tol, \ n = i, \ end \ for$

$$\tag{20}$$

Set $(T_n)_{i,i} = \alpha_i, \ (T_n)_{i+1,i} = (T_n)_{i,i+1} = \beta_i, otherwise \ (T_n)_{i,j} = 0$
$y_n = T_n^{-1/2} e_1^{(n)}/\rho_1 = U_n \Lambda_n^{-1/2} U_n^T e_1^{(n)}/\rho_1$

$q_0 = o, \ q_1 = \rho_1 b, \ x_0 = o$
for $i = 1, \ldots, n$
    $x_i = x_{i-1} + q_i y_n^{(i)}$
    $v = A^\dagger A q_i$
    $v := v - q_i \alpha_i - q_{i-1}\beta_{i-1}$
    $q_{i+1} = v/\beta_i$

where by o I denote a vector with zero entries and U_n, Λ_n the matrices of the egienvectors and eigenvalues of T_n. Note that there are only four large vectors necessary to store: q_{i-1}, q_i, v, x_i.

5   Testing the Method

I propose a simple test: I solve the system $A^\dagger A x = b$ by applying twice the *Algorithm2*, i.e. I solve the linear systems

$$(A^\dagger A)^{1/2} z = b, \quad (A^\dagger A)^{1/2} x = z \tag{21}$$

in the above order. For each approximation x_i, I compute the residual vector

$$r_i = b - A^\dagger A x_i \tag{22}$$

The method is tested for a SU(3) configuration at $\beta = 6.0$ on a $8^3 16$ lattice, corresponding to an order 98304 complex matrix A.

In Fig.1 I show the norm of the residual vector decreasing monotonically. The stagnation of $\|r_i\|_2$ for small values of *tol* may come from the accumulation of round off error in the 64-bit precision arithmetic used here.

This example shows that the tolerance line is above the residual norm line, which confirms the expectation that *tol* is a good stopping condition of the *Algorithm2*.

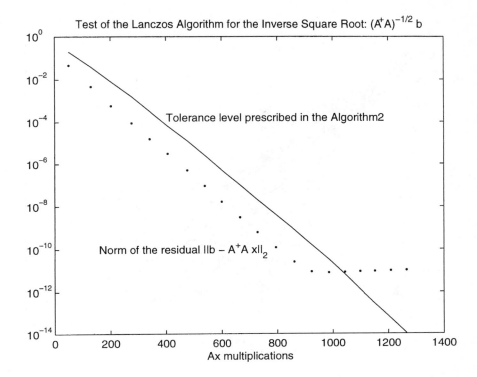

Fig. 1. The dots show the norm of the residual vector, whereas the line shows the tolerance level set by *tol* in the *Algorithm2*.

6   Inversion

Having computed the operator, one can invert it by applying iterative methods based on the the Lanczos algorithm. Since the operator D is normal, it turns out that the Conjugate Residual (CR) algorithm is the optimal one [10].

In Fig. 2 I show the converegence history of CR on 30 small 4^4 lattices at $\beta = 6$. The large number of multiplications with D_W suggests that the inversion of the Neuberger operator is a difficult task and may bring the complexity of quenched simulations in lattice QCD to the same order of magnitude to dynamical simulations with Wilson fermions. Therefore, other ideas are needed.

The essential point is the large number of small eigenvalues of A that make the computation of D time consuming. Therefore, one may try to project out these modes and invert them directly [11].

Also, one may try 5−dimensional implementations of the Neuberger operator, such that its condition improves [12].

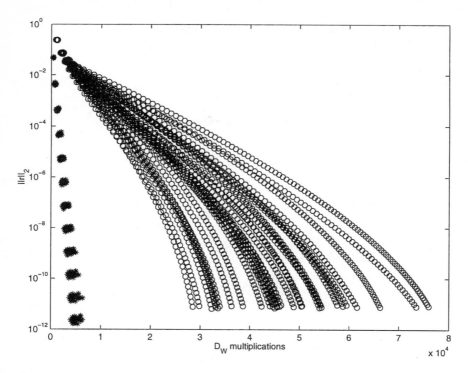

Fig. 2. Norm of the residual error vs. the number of D_W multiplications on 30 configurations. Circles stand for the straightforward inversion with CR and stars for the multigrid algorithm.

I have tried also to reformulate the theory in five dimensions by using the corresponding approximate inversion as a coarse grid solution in a multigrid scheme [13]. The scheme is tested and the results are shown in Fig. 2, where the multigrid pattern of the residual norm is clear. The gain with respect to CR is about a factor 10.

Note that to invert the "big" matrix I have used the BiCGstab2 algorithm [14] which is almost optimal in most of the cases for the non-normal matrices as is the matrix $\mathcal{M}$ [10].

7   Acknowledgement

The author would like to thank the organizers of this Workshop for the kind hospitality at Wuppertal.

References

1. P. H. Ginsparg and K. G. Wilson, Phys. Rev. D 25 (1982) 2649.

2. H. Neuberger, Phys. Lett. B 417 (1998) 141, Phys. Rev. D 57 (1998) 5417.
3. N. J. Higham, Proceedings of "Pure and Applied Linear Algebra: The New Generation", Pensacola, March 1993.
4. A. Boriçi, Phys.Lett. B453 (1999) 46-53, hep-lat/9910045
5. H. Neuberger, Phys. Rev. Lett. 81 (1998) 4060.
6. R. G. Edwards, U. M. Heller and R. Narayanan, Nucl.Phys. B540 (1999) 457-471.
7. B. Bunk, Nucl.Phys.Proc.Suppl. B63 (1998) 952.
8. P. Hernandes, K. Jansen, L. Lellouch, these proceedings and hep-lat/0001008.
9. G. H. Golub and C. F. Van Loan, *Matrix Computations*, The Johns Hopkins University Press, Baltimore, 1989. This is meant as a general reference with original references included therein.
10. A. Boriçi, *Krylov Subspace Methods in Lattice QCD*, PhD Thesis, CSCS TR-96-27, ETH Zürich 1996.
11. R. G. Edwards, U. M. Heller, J. Kiskis, R. Narayanan, hep-lat/9912042.
12. H. Neuberger, hep-lat/9909042
13. A. Boriçi, hep-lat/9907003, hep-lat/9909057
14. M. H. Gutknecht, SIAM J. Sci. Comput., 14 (1993) 1020.

On Lanczos-Type Methods for Wilson Fermions

Martin H. Gutknecht*

Seminar for Applied Mathematics
ETH Zürich, ETH-Zentrum
CH-8092 Zürich, Switzerland

Abstract. Numerical simulations of lattice gauge theories with fermions rely heavily on the iterative solution of huge sparse linear systems of equations. Due to short recurrences, which mean small memory requirement, Lanczos-type methods (including suitable versions of the conjugate gradient method when applicable) are best suited for this type of problem. The Wilson formulation of the lattice Dirac operator leads to a matrix with special symmetry properties that makes the application of the classical biconjugate gradient (BiCG) particularly attractive, but other methods, for example BiCGStab and BiCGStab2 have also been widely used. We discuss some of the pros and cons of these methods. In particular, we review the specific simplification of BiCG, clarify some details, and discuss general results on the roundoff behavior.

1 The Symmetry Properties of the Wilson Fermion Matrix

In the Wilson formulation of the lattice Dirac operator, where the Green's function of a single quark with bare mass m is computed by a model based on simple nearest neighbor coupling on a regular 4-dimensional space-time grid with periodic boundary conditions, the resulting linear system $\mathbf{W}\mathbf{x} = \mathbf{b}$ (which in lattice QCD is often written as $M\psi = \phi$) has a coefficient matrix of the form

$$\mathbf{W} = \mathbf{I} - \kappa\mathbf{B}, \tag{1}$$

where $\kappa \in \mathbb{R}$ is the so-called hopping parameter and $\mathbf{B}$ is a matrix of order $12 \times l_1 \times l_2 \times l_3 \times l_4$, with l_μ denoting the number of lattice points in dimension μ. Nowadays, typically $l_\mu = 16$, 32, or 64 for all μ, so that the order of the matrix ranges between $12 \times 16^4 = 786,432$ and $12 \times 64^4 \approx 2 \times 10^8$. The matrix $\mathbf{B}$ is well-known to have useful features [1,16,17]: first, $\mathbf{B}$ is formally Γ_5-Hermitian or Γ_5-selfadjoint, in the sense that[1]

$$\mathbf{B}^\star = \Gamma_5\mathbf{B}\Gamma_5, \qquad \text{where} \quad \Gamma_5 = \Gamma_5^\star = \Gamma_5^{-1} \tag{2}$$

---

* E-mail: mhg@sam.math.ethz.ch, www.sam.math.ethz.ch/~mhg
[1] The star denotes the adjoint or conjugate transpose of a matrix.

is a real diagonal matrix with elements ± 1, which takes the form

$$\Gamma_5 :\equiv \mathrm{diag}\left[\, 1\; 1\; \cdots\; 1\; -1\; -1\; \cdots\; -1\,\right]$$

if equations and unknowns are ordered appropriately; second, since the underlying discretization is restricted to nearest-neighbor coupling, $\mathbf{B}$ is at the same time "odd/even symmetric" in the sense that

$$\Sigma\mathbf{B} = -\mathbf{B}\Sigma, \tag{3}$$

where Σ is a diagonal matrix with $+1$'s and -1's. For example, in the two-dimensional case, a diagonal entry of Σ is $+1$ if for the corresponding grid point (i,j) the difference $i-j$ is even. This actually means that $\mathbf{B}$ is a so-called checker board matrix ("Schachbrett-Matrix" in German, see, $e.g.$, , [50]) with the property that $(\mathbf{B})_{k,l} = 0$ if $k-l$ is even. By suitable simultaneous row and column permutations that correspond to a red-black or even-odd reordering $\mathbf{B}$ can be brought into the form

$$\widetilde{\mathbf{B}} :\equiv \begin{bmatrix} \mathbf{O} & \widetilde{\mathbf{B}}_1 \\ \widetilde{\mathbf{B}}_2 & \mathbf{O} \end{bmatrix}, \tag{4}$$

which exhibits that $\widetilde{\mathbf{B}}$ is weakly 2-cyclic [47]. In general, $\mathbf{B}$ is a block checkerboard matrix, which can also be brought into the form (4).

The first symmetry, (2), implies that the spectrum of $\mathbf{B}$ is symmetric about the real axis, and the second, (3), entails that the spectrum is also symmetric about the origin, whence the spectrum is actually symmetric about both axes.

For the Wilson fermion matrix $\mathbf{W}$ we have due to (2)

$$\mathbf{W}^\star = \Gamma_5\mathbf{W}\Gamma_5, \tag{5}$$

and the spectrum is symmetric about the real axis and about the point 1. On the other hand, (4) implies that $\mathbf{W}$ has Young's Property A, which makes the linear system suitable for the SOR method [47], in particular since the spectrum is well captured by an ellipse whose larger axis covers only part of the interval $(0,2)$, as long as κ remains below a critical value. SOR can be expected to converge about twice as fast as the complex Chebyshev iteration [34,48], which is also an option, but does not take advantage of the (generalized) odd-even structure (4); see [22] for an analogous, but nonlinear problem from another application. SOR and the Chebyshev method require some preliminary knowledge about the spectrum (which may be obtained from a previous application of the biconjugate gradient method), but require no inner products, which is an important advantage on parallel computers.

The systems of the form $\mathbf{W}\mathbf{x} = \mathbf{b}$ that need to be solved are sometimes formally preconditioned with the matrix

$$\Sigma\mathbf{W}\Sigma = \mathbf{I} + \kappa\mathbf{B},$$

so that the system matrix becomes

$$(\varSigma \mathbf{W})^2 = \varSigma \mathbf{W} \varSigma \mathbf{W} = \mathbf{I} - \kappa^2 \mathbf{B}^2 \qquad (6)$$

and is seen to commute with $\varSigma$, hence, is a block checker board matrix of the other type (with non-zero block diagonal). This implies that the system decouples into two systems of half the size. Also this matrix and, hence, its two diagonal blocks of roughly half the size are $\varGamma_5$-adjoint. However, neither $\mathbf{W}$ nor $(\varSigma \mathbf{W})^2$ have a real spectrum.

In contrast, preconditioning by $\varGamma_5$ yields a linear system with the matrix $\varGamma_5 \mathbf{W}$, for which in view of (2) and $\kappa \in \mathbb{R}$

$$(\varGamma_5 \mathbf{W})^\star = (\varGamma_5 - \kappa \varGamma_5 \mathbf{B})^\star = \varGamma_5^\star - \kappa \mathbf{B}^\star \varGamma_5^\star = \varGamma_5 - \kappa \varGamma_5 \mathbf{B} = \varGamma_5 \mathbf{W}, \qquad (7)$$

which shows that $\varGamma_5 \mathbf{W}$ is Hermitian and, thus, has real spectrum.

Since the iterative solution of $\mathbf{W}\mathbf{x} = \mathbf{b}$ is so time and memory consuming, it is crucial to use algorithms that are particularly suitable for this special system and capitalize upon some of the special properties mentioned. We have referred to SOR in connection with the Property A and the special form of the spectrum of $\mathbf{W}$. Relation (7) suggests to apply a solver for Hermitian indefinite systems, such as MINRES, to $\varGamma_5 \mathbf{W}\mathbf{x} = \varGamma_5 \mathbf{b}$.

(However, we need to mention that a recent analysis of Sleijpen, van der Vorst, and Modersitzki [43] shows that the limiting accuracy of MINRES is far below that of most other methods.) Boriçi [1] and Frommer et al. [17] have made use of relation (2) for simplifying the biconjugate gradient (BICG) method, and we will discuss some not so well known details below. Experiments with these and other methods have been documented in many papers, see, for example, [1–3,7,14–16].

2   The Biconjugate Gradient Method and Some Related Methods

The first Lanczos-type method, introduced in 1952 by Lanczos [33] as the "complete algorithm for minimized iterations", is essentially what we now call the (standard) BIOMIN form [27] of the the biconjugate gradient (BICG) method [8]. It is fully analogous to the classical Hestenes-Stiefel version of the conjugate gradient (CG) method for Hermitian positive definite systems [30], which is also referred to as OMIN algorithm for CG. Unlike CG, which is restricted to Hermitian positive definite systems, BICG is applicable to general nonsingular square systems $\mathbf{A}\mathbf{x} = \mathbf{b}$. However, in contrast to CG, BICG may break down due to division by zero. If it does not, then, in exact arithmetic, BICG would converge in at most N steps, if N denotes the order of the system. In finite precision arithmetic, BICG is strongly influenced by roundoff errors and thus is not guaranteed to converge. But, in practice, when applied to very large systems, we anyway need methods that converge in much fewer than N steps.

Like the classical OMIN version of CG, BIOMIN is based on a pair of coupled recurrences for the residual and the direction polynomials. These recurrences are used to build up biorthogonal (or, dual) bases for a pair of nested sequences of dual Krylov spaces,

$$\mathcal{K}_n :\equiv \mathcal{K}_n(\mathbf{A}, \mathbf{y}_0) :\equiv \mathrm{span}\left(\mathbf{y}_0, \mathbf{A}\mathbf{y}_0, \ldots, \mathbf{A}^{n-1}\mathbf{y}_0\right), \tag{8}$$

$$\widetilde{\mathcal{K}}_n :\equiv \mathcal{K}_n(\mathbf{A}^\star, \widetilde{\mathbf{y}}_0) :\equiv \mathrm{span}\left(\widetilde{\mathbf{y}}_0, \mathbf{A}^\star\widetilde{\mathbf{y}}_0, \ldots, (\mathbf{A}^\star)^{n-1}\widetilde{\mathbf{y}}_0\right), \tag{9}$$

$n = 1, 2, \ldots$. The bases consist of the biorthogonal *Lanczos vectors* $\widetilde{\mathbf{y}}_m \in \widetilde{\mathcal{K}}_m$, $\mathbf{y}_n \in \mathcal{K}_n$ satisfying

$$\langle \widetilde{\mathbf{y}}_m, \mathbf{y}_n \rangle = \begin{cases} 0, & m \neq n, \\ \delta_n, & m = n. \end{cases} \tag{10}$$

At the same time another pair of bases is generated, consisting of the biconjugate *direction vectors* $\widetilde{\mathbf{v}}_m \in \widetilde{\mathcal{K}}_m$, $\mathbf{v}_n \in \mathcal{K}_n$ satisfying

$$\langle \widetilde{\mathbf{v}}_m, \mathbf{A}\mathbf{v}_n \rangle = \begin{cases} 0, & m \neq n, \\ \delta'_n, & m = n. \end{cases} \tag{11}$$

Additionally, approximations $\mathbf{x}_n \in \mathbf{x}_0 + \mathcal{K}_n$ of the solution of $\mathbf{A}\mathbf{x} = \mathbf{b}$ are computed, and in BICG, which is a Petrov–Galerkin method, these satisfy $\mathbf{b} - \mathbf{A}\mathbf{x}_n \perp \widetilde{\mathcal{K}}_n$. In view of (10), the (right-hand side) Lanczos vectors $\mathbf{y}_n$ also satisfy $\mathbf{y}_n \perp \widetilde{\mathcal{K}}_n$, and, in fact, they are normally scaled so that they coincide with the residuals, that is, $\mathbf{y}_n = \mathbf{r}_n :\equiv \mathbf{b} - \mathbf{A}\mathbf{x}_n$. Here is a summary of the resulting standard BICG algorithm[2].

Algorithm 1 (BIOMIN FORM OF THE BICG METHOD). For solving $\mathbf{A}\mathbf{x} = \mathbf{b}$ choose an initial approximation $\mathbf{x}_0$, set $\mathbf{v}_0 := \mathbf{y}_0 := \mathbf{b} - \mathbf{A}\mathbf{x}_0$, and choose $\widetilde{\mathbf{v}}_0 := \widetilde{\mathbf{y}}_0$ such that $\delta_0 := \langle \widetilde{\mathbf{y}}_0, \mathbf{y}_0 \rangle \neq 0$ and $\delta'_0 := \langle \widetilde{\mathbf{y}}_0, \mathbf{A}\mathbf{v}_0 \rangle \neq 0$. Then, for $n = 0, 1, \ldots$ compute

$$\omega_n := \delta_n / \delta'_n, \tag{12a}$$

$$\mathbf{y}_{n+1} := \mathbf{y}_n - \mathbf{A}\mathbf{v}_n\omega_n, \tag{12b}$$

$$\widetilde{\mathbf{y}}_{n+1} := \widetilde{\mathbf{y}}_n - \mathbf{A}^\star\widetilde{\mathbf{v}}_n\overline{\omega_n}, \tag{12c}$$

$$\mathbf{x}_{n+1} := \mathbf{x}_n + \mathbf{v}_n\omega_n, \tag{12d}$$

$$\delta_{n+1} := \langle \widetilde{\mathbf{y}}_{n+1}, \mathbf{y}_{n+1} \rangle, \tag{12e}$$

$$\psi_n := -\delta_{n+1} / \delta_n, \tag{12f}$$

$$\mathbf{v}_{n+1} := \mathbf{y}_{n+1} - \mathbf{v}_n\psi_n, \tag{12g}$$

$$\widetilde{\mathbf{v}}_{n+1} := \widetilde{\mathbf{y}}_{n+1} - \widetilde{\mathbf{v}}_n\overline{\psi_n}, \tag{12h}$$

$$\delta'_{n+1} := \langle \widetilde{\mathbf{v}}_{n+1}, \mathbf{A}\mathbf{v}_{n+1} \rangle. \tag{12i}$$

---

[2] The overbar denotes complex conjugation. Complex quantities can be avoided when all data $(\mathbf{A}, \mathbf{b}, \mathbf{x}_0)$ are real. We define the (complex) Euclidean inner product by $\langle \mathbf{z}, \mathbf{y} \rangle :\equiv \mathbf{z}^\star\mathbf{y} = \sum \overline{\zeta_k}\, \eta_k$.

If $\mathbf{y}_{n+1} \approx \mathbf{o}$, the process terminates and $\mathbf{x}_{n+1}$ is the solution; if $\delta_{n+1} \approx 0$ (and hence $\psi_n \approx 0$) or $\delta'_{n+1} \approx 0$, but $\mathbf{y}_{n+1} \not\approx \mathbf{o}$, the algorithm breaks down ("Lanczos and pivot breakdowns", respectively).

The recurrence coefficients ω_n and ψ_{n-1} are chosen so that the conditions (10) and (11) are satisfied for $m = n - 1$. The most important feature of BICG is that, in exact arithmetic, the other of these conditions are then satisfied automatically: the corresponding orthogonality is inherited — at least in exact arithmetic.

By eliminating the direction vectors from the recurrences of Algorithm 1 we obtain the BIORES form of the BICG method, where the Lanczos vectors are generated by three-term recurrences; see, *e.g.*, [27] for this connection, which is based on an LU decomposition of a tridiagonal matrix:

Algorithm 2 (BIORES FORM OF THE BICG METHOD). To solve $\mathbf{Ax} = \mathbf{b}$, choose an initial approximation $\mathbf{x}_0$, set $\mathbf{y}_0 := \mathbf{b} - \mathbf{Ax}_0$, and choose $\tilde{\mathbf{y}}_0$ such that $\delta_0 := \langle \tilde{\mathbf{y}}_0, \mathbf{y}_0 \rangle \neq 0$. Set $\beta_{-1} := 0$. Then, for $n = 0, 1, \ldots$ compute

$$\delta_n^{\mathbf{A}} := \langle \tilde{\mathbf{y}}_n, \mathbf{Ay}_n \rangle, \tag{13a}$$

$$\alpha_n := \delta_n^{\mathbf{A}} / \delta_n, \tag{13b}$$

$$\beta_{n-1} := \gamma_{n-1}\delta_n/\delta_{n-1} \qquad (\text{if} \quad n > 0), \tag{13c}$$

$$\gamma_n := -\alpha_n - \beta_{n-1}, \tag{13d}$$

$$\mathbf{y}_{n+1} := (\mathbf{Ay}_n - \mathbf{y}_n\alpha_n - \mathbf{y}_{n-1}\beta_{n-1})/\gamma_n, \tag{13e}$$

$$\tilde{\mathbf{y}}_{n+1} := (\mathbf{A}^\star\tilde{\mathbf{y}}_n - \tilde{\mathbf{y}}_n\overline{\alpha_n} - \tilde{\mathbf{y}}_{n-1}\overline{\beta_{n-1}})/\overline{\gamma_n}, \tag{13f}$$

$$\delta_{n+1} := \langle \tilde{\mathbf{y}}_{n+1}, \mathbf{y}_{n+1} \rangle, \tag{13g}$$

$$\mathbf{x}_{n+1} := -(\mathbf{y}_n + \mathbf{x}_n\alpha_n + \mathbf{x}_{n-1}\beta_{n-1})/\gamma_n. \tag{13h}$$

If $\gamma_n \approx 0$, the algorithm breaks down ("pivot breakdown"). If $\mathbf{y}_{n+1} \approx \mathbf{o}$, it terminates and $\mathbf{x}_{n+1}$ is the solution. If $\mathbf{y}_{n+1} \not\approx \mathbf{o}$, but $\delta_{n+1} \approx 0$, it also breaks down ("Lanczos breakdown").

A serious shortcoming of BICG and related, so-called Lanczos-type methods for nonsymmetric systems is the possibility of breakdowns. These were probably the main reason why for decades numerical analysts were very reluctant to apply or even promote this method. Finally, look-ahead steps were introduced to circumnavigate such breakdowns [38,24,26,11]. It was also noticed that in practice breakdowns and near-breakdowns with serious effects are quite rare. Ever since, BICG and other Lanczos-type methods have become more and more popular, although look-ahead is rarely implemented.

Another disadvantage is that in contrast to most other Krylov space methods BICG requires two matrix-vector products (MVs) per step, but only increases the search space $\mathcal{K}_n$ by one dimension. This disadvantages of BICG was overcome by Sonneveld [44] with the introduction of the *conjugate gradient squared* (CGS) *method*, which should rather be called BICGS and will be referred to here as (BI)CGS. Sonneveld's clever idea was to derive recurrences that produce approximations $\mathbf{x}_n \in \mathbf{x}_0 + \mathcal{K}_{2n}$ whose residuals $\mathbf{r}_n \in \mathcal{K}_{2n+1}$ correspond to the squares p_n^2 of the residual polynomials p of BiCG (which are often referred to as the Lanczos polynomials). If we denote by $\widehat{p}_n$ the polynomials that are associated with the direction vectors $\mathbf{v}_n$ of BiCG, then these recurrences involve, in addition to the iterates $\mathbf{x}_n$, the vector sequences

$$\mathbf{r}_n := p_n^2(\mathbf{A})\mathbf{r}_0 \in \mathcal{K}_{2n+1}, \qquad\qquad \mathbf{s}_n := p_n(\mathbf{A})\widehat{p}_n(\mathbf{A})\mathbf{r}_0 \in \mathcal{K}_{2n+1},$$

$$\mathbf{s}'_n := p_{n+1}(\mathbf{A})\widehat{p}_n(\mathbf{A})\mathbf{r}_0 \in \mathcal{K}_{2n+2}, \qquad \widehat{\mathbf{r}}_n := \widehat{p}_n^2(\mathbf{A})\mathbf{r}_0 \in \mathcal{K}_{2n+1}.$$

They are easily derived from the BiOMin recurrences (12b)–(12c) and (12g)–(12h), and they contain the same coefficients $\omega_n := \delta_n/\delta'_n$ and $\psi_n := -\delta_{n+1}/\delta_n$. Note that

$$\delta_n = \langle \widetilde{\mathbf{y}}_n, \mathbf{y}_n \rangle = \langle \overline{p_n}(\mathbf{A}^\star)\widetilde{\mathbf{y}}_0, p_n(\mathbf{A})\mathbf{r}_0 \rangle = \langle \widetilde{\mathbf{y}}_0, p_n^2(\mathbf{A})\mathbf{r}_0 \rangle = \langle \widetilde{\mathbf{y}}_0, \mathbf{r}_n \rangle,$$

$$\delta'_n = \langle \widetilde{\mathbf{v}}_n, \mathbf{A}\mathbf{v}_n \rangle = \langle \overline{\widehat{p}_n}(\mathbf{A}^\star)\widetilde{\mathbf{v}}_0, \mathbf{A}\widehat{p}_n(\mathbf{A})\mathbf{r}_0 \rangle = \langle \widetilde{\mathbf{y}}_0, \widehat{p}_n^2(\mathbf{A})\mathbf{A}\mathbf{r}_0 \rangle = \langle \widetilde{\mathbf{y}}_0, \mathbf{A}\widehat{\mathbf{r}}_n \rangle. \tag{14}$$

A typical behavior of BiCG is that the residual norms $\|\mathbf{y}_n\|$ fluctuate strongly, in particular when the problem solved is ill conditioned and, consequently, the convergence is rather slow. In (Bi)CGS this erratic convergence behavior is even more pronounced. One way to counteract it is by replacing the residual polynomials p_n^2 of (Bi)CGS by a more general product $p_n t_n$, where t_n belongs to another polynomial sequence satisfying a short recurrence. This leads to *Lanczos-type product methods* (*LTPMs*). The first algorithm of this class was BiCGStab, due to van der Vorst [45], where t_n is built up from linear factors: $t_{n+1}(\zeta) = (1 - \chi_{n+1}\zeta)t_n(\zeta)$, and where the sequences

$$\mathbf{r}_n :\equiv p_n(\mathbf{A})t_n(\mathbf{A})\mathbf{r}_0, \quad \widehat{\mathbf{r}}_n :\equiv \widehat{p}_n(\mathbf{A})t_n(\mathbf{A})\mathbf{r}_0, \quad \mathbf{w}_n :\equiv p_{n+1}(\mathbf{A})t_n(\mathbf{A})\mathbf{r}_0,$$

are constructed in addition to the iterates $\mathbf{x}_n$. The coefficient χ_{n+1} is chosen such that

$$\|\mathbf{r}_{n+1}\| = \min_{\chi} \|\mathbf{w}_n - \mathbf{A}\mathbf{w}_n\chi\|.$$

For BiCGStab the residual norm history is typically much smoother than for (Bi)CGS, but a disadvantage of this method is that the zeros $1/\chi_n$ of the second set $\{t_n\}$ of polynomials are necessarily all real when a real-valued problem is solved in real arithmetic, even when the spectrum of the matrix is truly complex. Moreover, they remain fixed for all subsequent polynomials of the set. The first disadvantage is avoided if the linear factors are replaced by quadratic factors that are attached every other step; they allow a two-dimensional residual minimization in every other step, as suggested in BiCG-Stab2 [25]. The second disadvantage is overcome if the second set is chosen to satisfy a three-term recurrence or a pair of coupled two-term recurrences (which can be used for a two-dimensional residual minimization in every step), as suggested by Zhang in his GPBI-CG algorithm [49] (an equivalent form of which is called BiCG×MR2 in [27]).

3 Simplifications due to Symmetries

When $\mathbf{A}$ is Hermitian, the choice $\widetilde{\mathbf{y}}_0 := \mathbf{y}_0$ will produce in the BiORes algorithm identical left and right vectors, $\widetilde{\mathbf{y}}_n = \mathbf{y}_n$ ($\forall n$), and in the BiOMin algorithm also $\widetilde{\mathbf{v}}_n = \mathbf{v}_n$ ($\forall n$), so that there is no need to compute the left sequences $\{\widetilde{\mathbf{y}}_n\}$ and $\{\widetilde{\mathbf{v}}_n\}$. This is easily seen by induction and by noting that the numbers α_n, β_n, γ_n, ψ_n, and ω_n are real, even when $\mathbf{A}$ or $\mathbf{y}_0$ are complex. The resulting simplified algorithms are then exactly the OMin and the ORes algorithms, respectively, for CG; they may be applied also to indefinite Hermitian systems, but then they can break down too.

One may raise the question whether there are other situations where BiCG simplifies in the sense that only one MV is required per step. In 1953, Rutishauser [39] and later Fletcher [8], both assuming real data, pointed out that in the three-term Lanczos process (and thus also in BiORes, which just makes use of the special normalization $\gamma_n := -\alpha_n - \beta_{n-1}$), the knowledge of a matrix $\mathbf{S}$ satisfying

$$\mathbf{A}^\top = \mathbf{S}\mathbf{A}\mathbf{S}^{-1} \tag{15}$$

allows us to make such a reduction: choosing $\widetilde{\mathbf{y}}_0 := \mathbf{S}\mathbf{y}_0$ yields $\widetilde{\mathbf{y}}_n := \mathbf{S}\mathbf{y}_n$ ($n > 0$). In [23] we mentioned this again and made the simple observation that the complex case is covered too when we choose $\widetilde{\mathbf{y}}_0 := \overline{\mathbf{S}\mathbf{y}_0}$, which then yields $\widetilde{\mathbf{y}}_n := \overline{\mathbf{S}\mathbf{y}_n}$ ($n > 0$), as is readily verified. Thus we can delete (13f) if we replace (13a) and (13g) by

$$\delta_n^{\mathbf{A}} := \langle \overline{\mathbf{y}_n}, \mathbf{A}\mathbf{y}_n \rangle_{\mathbf{S}^\top} , \tag{16}$$

$$\delta_{n+1} := \langle \overline{\mathbf{y}_{n+1}}, \mathbf{y}_{n+1} \rangle_{\mathbf{S}^\top} , \tag{17}$$

respectively, where

$$\langle \mathbf{z}, \mathbf{y} \rangle_{\mathbf{S}^\top} :\equiv \langle \mathbf{z}, \mathbf{S}^\top \mathbf{y} \rangle = \mathbf{z}^\star \mathbf{S}^\top \mathbf{y} . \tag{18}$$

The BiOMin algorithm simplifies in a fully analogous way: we just need additionally

$$\delta'_{n+1} := \langle \overline{\mathbf{v}_{n+1}}, \mathbf{A}\mathbf{v}_{n+1} \rangle_{\mathbf{S}^\top} \tag{19}$$

in order to delete (12c) and (12h). Rutishauser also pointed out that a matrix $\mathbf{S}$ satisfying (15) always exists, as every matrix is known to be similar to its transposed, but the usual proof for this makes use of the Jordan canonical form, see, *e.g.*, [32, p. 134]. Of course, the simplification is only useful, if $\mathbf{S}$ is known and if the matrix-vector products $\mathbf{S}\mathbf{y}_n$ are cheaper than $\mathbf{A}^\star \widetilde{\mathbf{y}}_n$.

Note that (15) does not include the simple Hermitian case $\mathbf{A}^\star = \mathbf{A}$, but it covers the complex symmetric case (where $\mathbf{S} = \mathbf{I}$), which was treated in detail by Freund [9].

In [10] Freund looked for further situations where the Lanczos process simplifies, and in particular for classes of matrices where the matrix $\mathbf{S}$ is known due to the special structure of $\mathbf{A}$. For example, for a Toeplitz matrix (15) holds with $\mathbf{S}$ the antidiagonal unit matrix $\mathbf{J}$. More generally, a matrix satisfying (15) with $\mathbf{S} = \mathbf{J}$ is symmetric about the antidiagonal and is called persymmetric. In [10] Freund treated the cases

$$\mathbf{A}^{\top} = \mathbf{S}\mathbf{A}\mathbf{S}^{-1}, \qquad \mathbf{S} = \mathbf{S}^{\top}, \tag{20}$$

and

$$\mathbf{A}^{\star} = \mathbf{S}\mathbf{A}\mathbf{S}^{-1}, \qquad \mathbf{S} = \mathbf{S}^{\star}. \tag{21}$$

Clearly, (20) is a special case of (15) (the extra condition $\mathbf{S} = \mathbf{S}^{\top}$ is not needed for the simplification), but for complex matrices (21) is different. In [13], Freund and Nachtigal then referred to the two cases (15) and

$$\mathbf{A}^{\star} = \mathbf{S}\mathbf{A}\mathbf{S}^{-1}, \tag{22}$$

that is, they dropped the symmetry assumption for $\mathbf{S}$ in (21), which, however, seems to be wrong[3]. In fact, the recipe is to choose[4] $\widetilde{\mathbf{y}}_0 := \mathbf{S}\mathbf{y}_0$ in BiCG and to aim for $\widetilde{\mathbf{y}}_n = \mathbf{S}\mathbf{y}_n$ $(n > 0)$. Inserting this and (22) in (13f) leads after premultiplication with $\mathbf{S}^{-1}$ to

$$\mathbf{y}_{n+1} := (\mathbf{A}\mathbf{y}_n - \mathbf{y}_n\overline{\alpha_n} - \mathbf{y}_{n-1}\overline{\beta_{n-1}})/\overline{\gamma_n}, \tag{23}$$

which differs from (13e) only in the complex conjugated coefficients. By making additionally use of $\mathbf{S}^{\star} = \mathbf{S}$, we see that (21) implies that

$$(\mathbf{S}\mathbf{A})^{\star} = \mathbf{S}\mathbf{A}, \tag{24}$$

that is $\mathbf{S}\mathbf{A}$ is Hermitian. Consequently,

$$\delta_n :\equiv \langle \widetilde{\mathbf{y}}_n, \mathbf{y}_n \rangle = \langle \mathbf{y}_n, \mathbf{S}\mathbf{y}_n \rangle \in \mathbb{R}, \tag{25}$$

$$\delta_n^{\mathbf{A}} :\equiv \langle \widetilde{\mathbf{y}}_n, \mathbf{A}\mathbf{y}_n \rangle = \langle \mathbf{y}_n, \mathbf{S}\mathbf{A}\mathbf{y}_n \rangle \in \mathbb{R}, \tag{26}$$

so that $\alpha_n \in \mathbb{R}$, $\beta_{n-1} \in \mathbb{R}$, and $\gamma_n \in \mathbb{R}$. Therefore, under the assumption (21) the BiORes algorithm can indeed be simplified, and the same is true for BiOMin since, when $\widetilde{\mathbf{v}}_n := \mathbf{S}\mathbf{v}_n$ also

$$\delta_n' :\equiv \langle \widetilde{\mathbf{v}}_n, \mathbf{v}_n \rangle = \langle \mathbf{v}_n, \mathbf{S}\mathbf{A}\mathbf{v}_n \rangle \in \mathbb{R}, \tag{27}$$

so that $\psi_n \in \mathbb{R}$ and $\omega_n \in \mathbb{R}$. This can all be recast in a proof by induction showing that choosing $\widetilde{\mathbf{y}}_0 := \mathbf{S}\mathbf{y}_0 := \mathbf{S}\mathbf{y}_0$ yields $\widetilde{\mathbf{y}}_n = \mathbf{S}\mathbf{y}_n$ for $n > 0$ in

---

[3] We must admit that we made the same mistake in a remark in §6.1 of [27], where we moreover claimed incorrectly that (21) implies that the spectrum of $\mathbf{A}$ is real.

[4] Note that Freund uses in the complex Lanczos process a formal, bilinear inner product $\mathbf{w}^{\top}\mathbf{y}$ instead of the usual sesquilinear inner product $\langle \widetilde{\mathbf{y}}, \mathbf{y} \rangle = \widetilde{\mathbf{y}}^{\star}\mathbf{y}$; therefore, up to a scalar factor, his left Lanczos vectors $\mathbf{w}_n$ and ours are related by $\widetilde{\mathbf{y}}_n = \overline{\mathbf{w}_n}$.

BiORes, and likewise, additionally choosing $\tilde{\mathbf{v}}_0 := \mathbf{S}\mathbf{v}_0 := \mathbf{S}\mathbf{y}_0$ in BiOMin implies $\tilde{\mathbf{v}}_n = \mathbf{S}\mathbf{v}_n$ for $n > 0$. In summary, for simplifying BiORes and BiOMin when (21) holds, we redefine δ_n, $\delta_n^{\mathbf{A}}$, and δ_n' as given in (25), (26), and (27) in order to delete (13f) in BiORes and (12c), (12h) in BiOMin, as has been proposed by Boriçi [1] and Frommer et al. [17] for the Wilson fermion computations.

Without the condition, $\mathbf{S} = \mathbf{S}^\star$, that is, assuming (22) alone, does not seem to lead to such a simplification, even if we turn to the most general versions of the Lanczos process [27] where γ_n and $\overline{\gamma_n}$ can be chosen freely (the latter need not be the complex conjugate of the former).

Software for the so simplified BiCG algorithms (and of the related QMR algorithm that is not discussed here) is available from

```
http://www.math.uni-wuppertal.de/org/SciComp/Projects/QCD.html
```

4   Finite Precision Effects

Roundoff errors can have strong effects on Lanczos-type methods (including CG). This is first of all due to the fact that the methods rely essentially on a variation of the Gram-Schmidt process, which is known to be prone to roundoff effects. Second, particularly in the nonsymmetric case, the computed recurrence coefficients may turn out to have large relative error. Third, the residuals are normally updated using recurrences and, hence, may differ considerably from the true residuals of the approximations $\mathbf{x}_n$. We will now discuss these three types of finite precision effects.

4.1   Loss of Orthogonality and Loss of Linear Independence

Recall that, for example, in CG and BiCG a Gram-Schmidt process is applied to make the residual $\mathbf{y}_{n+1}$ orthogonal to the earlier ones or the earlier left Lanczos vectors, respectively. The Gram-Schmidt process makes vectors shorter due to the subtraction of certain projections, and thus tiny errors in the coefficients or in the computation of the linear combinations may ultimately cause large relative errors, and, in particular, a *loss of orthogonality*: $\mathbf{y}_{n+1}$ will not be exactly orthogonal to $\tilde{\mathbf{y}}_0, \ldots, \tilde{\mathbf{y}}_n$. This loss of orthogonality is often severe enough to lead to a *loss of linear independence*.

Special is here that it suffices to enforce the orthogonality to $\tilde{\mathbf{y}}_n$ and $\tilde{\mathbf{y}}_{n-1}$ (in the case of BiCG), because the orthogonality to $\tilde{\mathbf{y}}_0, \ldots, \tilde{\mathbf{y}}_{n-2}$ is inherited in exact arithmetic. This has the advantage that there are only two subtractions, hence the resulting vector $\mathbf{y}_{n+1}\gamma_n$ will not be so much shorter than the one we started with, $\mathbf{A}\mathbf{y}_n$, but the drawback is that the loss of orthogonality may be worse since previous errors are inherited too. Hence, often

$$\frac{\langle \tilde{\mathbf{y}}_m, \mathbf{y}_n \rangle}{\|\tilde{\mathbf{y}}_m\| \, \|\mathbf{y}_n\|} \not\approx 0 \quad \text{if} \quad |m - n| \quad \text{large.}$$

For example, the fraction can be easily on the order of 10^{-1}.

This loss of orthogonality is particularly annoying when the tridiagonal matrix $\mathbf{T}_n$ generated in the Lanczos process is used to find approximate eigenvalues of $\mathbf{A}$, because it will cause $\mathbf{T}_n$ to have multiple copies of some of these eigenvalues. Full reorthogonalization, that is, repeating the Gram-Schmidt process with respect to the full set of Lanczos vectors (instead of the last two) would help, but the cost forbids this, since it would be necessary to store all Lanczos vectors. Two strategies have been developed to cope with this difficulty: either the so-called ghost eigenvalues are identified and removed as proposed by Cullum and Willoughby [4], or their creation is avoided by reducing the roundoff errors of the Lanczos vectors, as suggested by Parlett and his group [36,37,6,5]. However, it does not suffice to reduce the roundoff in the three-term Gram-Schmidt process (where $\mathbf{y}_{n+1}$ is made orthogonal to $\widetilde{\mathbf{y}}_n$ and $\widetilde{\mathbf{y}}_{n-1}$) by applying modified Gram-Schmidt or repeated classical Gram-Schmidt (there is little benefit because there are only three terms). Additionally, $\mathbf{y}_{n+1}$ needs to be reorthogonalized with respect to a selection of earlier Lanczos vectors. In the symmetric case, where this technique was explored first, this *selective reorthogonalizition* can be justified by Paige's roundoff analysis for the symmetric Lanczos process. The non-symmetric case was later treated by Day [6,5], who systematically explored measures for *maintaining duality*, that is, biorthogonality.

Since selective reorthogonalizition increases the program complexity and the memory requirements, computational physicists tend to prefer the Cullum and Willoughby filtering.

We should mention, however, that even if $\mathbf{T}_n$ is affected by large roundoff errors occuring in the Lanczos process, the implications are not completely devastating: groups of ghost eigenvalues somehow maintain the projection properties of the operator.

For solving linear systems it seems not really worth-while to apply all these tricks. Moreover, it seems to be impossible to adapt them to LTPMs, which are now considered to be the most effective solvers. In fact, when linear systems are solved, the loss of linear independence caused by a loss of orthogonality will just entail a *slowdown of the convergence*. In the symmetric case, this mechanism is well understood due to the work of Greenbaum and Strakoš [18,21]: *in finite precision arithmetic, the Lanczos process behaves like one for a bigger problem in exact arithmetic.*

4.2   Inaccurate Recurrence Coefficients and near-Breakdowns

To some extent, inaccurate recurrence coefficients in CG and BiCG are clearly linked to the loss of orthogonality just discussed. One effect adds to the other: inaccurate Lanczos vectors lead to inaccurate coefficients, and vice versa. However, in the non-Hermitian case, there is the additional danger that the inner products δ_n and δ'_n may be very small even if the vectors they are formed from are not short. Since an inner product of nearly orthogonal vectors is inherently prone to large relative roundoff error, these cases are dangerous: the quantities computed from δ_n and δ'_n will also have large error. This is then called a *near-breakdown*, since, when one of these inner products is needed and turns out to be exactly 0, then the corresponding algorithm breaks down due to a division by zero (*exact breakdown*). We refer to the case $\delta_n \approx 0$ as *Lanczos breakdown*, and to $\delta'_n \approx 0$ as *pivot breakdown*. In the Hermitian indefinite case (where $\widetilde{\mathbf{y}}_n = \mathbf{y}_n$) Lanczos breakdowns cannot occur, but pivot breakdowns still can. In the BIORES version of BICG and in the ORES version of CG (when applied to a Hermitian indefinite system), the pivot breakdown reappears as $\gamma_n \approx 0$. Only in the case of a Hermitian positive definite (Hpd) system, CG cannot break down.

Specifically, in BIORES the recurrence coefficients α_n and/or β_{n-1} are inaccurate if any of the inner products

$$\delta_n :\equiv \langle \widetilde{\mathbf{y}}_n, \mathbf{y}_n \rangle, \qquad \delta_{n-1} :\equiv \langle \widetilde{\mathbf{y}}_{n-1}, \mathbf{y}_{n-1} \rangle, \qquad \delta_n^{\mathbf{A}} :\equiv \langle \widetilde{\mathbf{y}}_n, \mathbf{A}\mathbf{y}_n \rangle$$

has large relative error. Moreover, $\gamma_n :\equiv -\alpha_n - \beta_{n-1}$ may be inaccurate if $|\gamma_n| \ll |\alpha_n|$.

Similarly, in BIOMIN, ψ_{n-1} and/or ω_n are inaccurate if any of the inner products

$$\delta_n :\equiv \langle \widetilde{\mathbf{y}}_n, \mathbf{y}_n \rangle, \qquad \delta_{n-1} :\equiv \langle \widetilde{\mathbf{y}}_{n-1}, \mathbf{y}_{n-1} \rangle, \qquad \delta'_n :\equiv \langle \widetilde{\mathbf{v}}_n, \mathbf{A}\mathbf{v}_n \rangle$$

has large relative error. And likewise, the same types of inaccuracy occur in (BI)CGS if any of the inner products

$$\delta_n :\equiv \langle \widetilde{\mathbf{y}}_0, \mathbf{r}_n \rangle, \qquad \delta_{n-1} :\equiv \langle \widetilde{\mathbf{y}}_0, \mathbf{r}_{n-1} \rangle, \qquad \delta'_n :\equiv \langle \widetilde{\mathbf{y}}_0, \mathbf{A}\widehat{\mathbf{r}}_n \rangle$$

has large relative error; see (14).

Except for $\delta_n^{\mathbf{A}} \approx 0$, all of these cases cause near-breakdowns. In particular, a pivot near-breakdown ($\gamma_n \approx 0$ in BIORES or $\delta'_n \approx 0$ in BIOMIN) causes not only large local errors in coefficients and vectors, but also very large vectors $\mathbf{x}_n$ and $\mathbf{r}_n$.

Fortunately, in linear system solvers, inaccurate recurrence coefficients only seem to have a strong effect on the convergence when the coefficients are very inaccurate, as it may happen when a near-breakdown occurs. As we mentioned before, for computing approximate eigenvalues the situation is different.

The bad effects of a breakdown or near-breakdown can be avoided by switching to look-ahead steps when necessary, a technique that was developed in [38] for eigenvalue computations, in [24,26,11,12] for various versions of BICG, and in [28] for LTPMs. Additional contributions and alternative approaches are referred to in [27] and [28], where also a simplification due to Hochbruck [31] is covered. In particular, divisions by near-zeros can be avoided by look-ahead, but there are still some open questions regarding the best strategy for its application.

Further possibilities to improve the accuracy of the recurrence coefficients include, in addition to those for reducing the loss of orthogonality mentioned in the previous subsection:

(i) The application of multiple precision arithmetic to compute the above inner products and the sum $\gamma_n :\equiv -\alpha_n - \beta_{n-1}$, an option one tries to avoid.

(ii) Alternative choices for the left vectors in BiCG. It is well-known (see Algorithm 3 in Saad [40] and Section 6.2 in [27]) that in BiCG the left vectors $\widetilde{\mathbf{y}}_n$ need not be chosen as Lanczos vectors, but could come from another nested basis for the dual space. In fact, LPTMs capitalize exactly upon this freedom. However, experiments done independently by Miroslav Rozložník (private communication) and the author have not turned out a convincing choice different from the standard one. Indeed the theory provides little hope for success in this way.

(iii) Replacing (Bi)CGS by an LTPM with suitably chosen second set of polynomials t_n. However, again there is limited hope for a strong improvement.

4.3   The Gap Between Updated and True Residuals

In most Krylov space methods one has the option to compute the residuals $\mathbf{r}_n :\equiv \mathbf{b} - \mathbf{A}\mathbf{x}_n$ explicitly according to this definition or by updating, that is by using some recursion(s) like the coupled two-term recursions (12b), (12g) of BiOMin and the three-term recursion (13e) of BiORes (recall that in BiCG $\mathbf{r}_n = \mathbf{y}_n$). In some cases, the explicit evaluation costs an extra matrix-vector product (MV), but normally another one can be avoided instead. Nevertheless, the folklore is that updating should be used because explicit computation adds to the roundoff in the process of generating the Krylov space, that is, produces a less accurate basis, and thus often slows down convergence. We therefore assume here that the residuals are computed by updating, and we let $\mathbf{r}_n$ denote the nth residual vector so obtained in finite precision arithmetic. Likewise, $\mathbf{x}_n$ is now the iterate computed in finite precision arithmetic, and $\mathbf{b} - \mathbf{A}\mathbf{x}_n$ is the *true residual* obtained in exact arithmetic from $\mathbf{x}_n$. (Actually, in numerical experiments the true residuals are computed in finite precision arithmetic too, but the error in the evaluation of this expression will normally be considerably smaller than the true residual itself, and this is all that is needed in this context.) Clearly, a *gap*

$$\mathbf{f} :\equiv \mathbf{b} - \mathbf{A}\mathbf{x}_n - \mathbf{r}_n \tag{28}$$

between the true and the updated residuals will occur, and one can expect that it will somehow grow with n. This has been known for a long time, but only recently this gap was analyzed for the two most important cases, namely for two-term update formulas

$$\begin{aligned} \mathbf{r}_{n+1} &:= \mathbf{r}_n - \mathbf{A}\mathbf{v}_n\omega_n\,, \\ \mathbf{x}_{n+1} &:= \mathbf{x}_n + \mathbf{v}_n\omega_n \end{aligned} \tag{29}$$

(which need to be combined with one for the direction vectors, say, $\mathbf{v}_0 := \mathbf{r}_0$, $\mathbf{v}_n := \mathbf{r}_n + \mathbf{v}_{n-1}\psi_{n-1}$ $(n > 0)$, which has no influence on the gap) like in BiOMin and in the classical OMin version of CG, and for a pair of three-term recurrences

$$\left.\begin{array}{l} \mathbf{r}_{n+1} := (\mathbf{A}\mathbf{r}_n - \mathbf{r}_n\alpha_n - \mathbf{r}_{n-1}\beta_{n-1})/\gamma_n, \\ \mathbf{x}_{n+1} := -(\mathbf{r}_n + \mathbf{x}_n\alpha_n + \mathbf{x}_{n-1}\beta_{n-1})/\gamma_n \end{array}\right\} \quad \text{with} \quad \gamma_n := -(\alpha_n + \beta_{n-1}),$$
$$(30)$$

like in BiORes and the corresponding ORes version of CG. (At the start, $\mathbf{r}_0 := \mathbf{b} - \mathbf{A}\mathbf{x}_0$, $\mathbf{r}_{-1} := \mathbf{o}$, $\mathbf{x}_{-1} := \mathbf{o}$, $\beta_{-1} := 0$.)

The relevance of this gap is due to the fact that in most methods the updated residuals become ultimately orders of magnitude smaller than the true residuals, which essentially stagnate from a certain moment. Consequently, a large gap means low attainable accuracy: the true residuals will stagnate early.

For the two-term recurrences of the form (29) Greenbaum [19,20] proved the following result (which improves a similar one of Sleijpen, van der Vorst, and Fokkema [42]):

Theorem 3. *Assume iterates and residuals are updated according to (29). Then the gap (28) between the true and the updated residual is given by*

$$\mathbf{f}_n = \mathbf{f}_0 - \mathbf{l}_0 - \cdots - \mathbf{l}_n, \qquad (31)$$

where

$$\mathbf{l}_n :\equiv \mathbf{A}\mathbf{h}_n + \mathbf{g}_n \qquad (32)$$

is the local error whose components $\mathbf{h}_n$ and $\mathbf{g}_n$ are defined by

$$\mathbf{x}_{n+1} = \mathbf{x}_n + \mathbf{v}_n\omega_n + \mathbf{h}_n, \qquad \mathbf{r}_{n+1} = \mathbf{r}_n - \mathbf{A}\mathbf{v}_n\omega_n + \mathbf{g}_n. \qquad (33)$$

In particular,

$$\frac{||\mathbf{f}_n||}{||\mathbf{A}||\,||\mathbf{x}||} \leq (\epsilon + \mathcal{O}(\epsilon^2))\,[n + 2 + (1 + \mu + (n+1)(10 + 2\mu))\Theta_n], \qquad (34)$$

where ϵ denotes the machine-epsilon, $\mu :\equiv m\sqrt{N}$ with m the maximum number of nonzeros in a row of $\mathbf{A}$ and N the matrix order, and

$$\Theta_n :\equiv \max_{k \leq n} \frac{||\mathbf{x}_k||}{||\mathbf{x}||}. \qquad (35)$$

In contrast, for a pair of three-term recurrences (30) the following holds [29]:

Theorem 4. *Assume iterates and residuals are updated according to* (30). *Then the gap* (28) *satisfies, up to* $\mathcal{O}(\epsilon^2)$,

$$
\begin{aligned}
\mathbf{f}_{n+1} = \mathbf{f}_0 \quad & - \; \mathbf{l}_0 \\
& - \; \mathbf{l}_0 \frac{\beta_0}{\gamma_1} - \mathbf{l}_1 \\
& - \; \mathbf{l}_0 \frac{\beta_0 \beta_1}{\gamma_1 \gamma_2} - \mathbf{l}_1 \frac{\beta_1}{\gamma_2} - \mathbf{l}_2 \\
& \;\; \vdots \\
& - \; \mathbf{l}_0 \frac{\beta_0 \beta_1 \cdots \beta_{n-1}}{\gamma_1 \gamma_2 \cdots \gamma_n} - \cdots - \mathbf{l}_{n-1} \frac{\beta_{n-1}}{\gamma_n} - \mathbf{l}_n \,,
\end{aligned}
\tag{36}
$$

where

$$
\mathbf{l}_n := \equiv (-\mathbf{b}\varepsilon_n + \mathbf{A}\mathbf{h}_n + \mathbf{g}_n)/\gamma_n
\tag{37}
$$

is the local error whose components $\mathbf{h}_n$, $\mathbf{g}_n$, *and* ε_n *are defined by*

$$
\begin{aligned}
\mathbf{r}_{n+1} &= (\mathbf{A}\mathbf{r}_n - \mathbf{r}_n \alpha_n - \mathbf{r}_{n-1} \beta_{n-1} + \mathbf{g}_n)/\gamma_n \,, \\
\mathbf{x}_{n+1} &= -(\mathbf{r}_n + \mathbf{x}_n \alpha_n + \mathbf{x}_{n-1} \beta_{n-1} + \mathbf{h}_n)/\gamma_n \,, \\
\gamma_n &= -(\alpha_n + \beta_{n-1} + \varepsilon_n).
\end{aligned}
\tag{38}
$$

It is rather easy to derive from the definitions of the local errors, that is, from (32), (33) and (37), (38), respectively, bounds for these local errors. They show that typically the local errors in algorithms based on three-term updates are larger than those arising in two-term updates. In the former case, (34) and (35) show that the size of the gap mainly depends on the norm of the largest iterate. In the latter case, the largest residual norm also has a direct influence, and the constants in the estimate are larger.

However, the main difference between Theorems 3 and 4 lies in the explicit formulas (31) and (36) for the gaps: while in the two-term case the gap $\mathbf{f}_n$ is just a sum of local errors $\mathbf{l}_j$, in the three-term case the (normally larger) local errors are multiplied (and thus amplified) by potentially very large factors. So, the gap is typically much bigger in the latter case. This leads to an explanation of the fact that the attainable accuracy, that is, the level on which the true residuals stagnate, is much worse for a three-term based algorithm than for one using two-term updates. This fact can be easily verified numerically in examples where the residual norm fluctuates heavily, a quite common behavior when $\mathbf{A}$ is ill-conditioned. This behavior is more likely in BiCG and other Lanczos-type methods for non-Hermitian systems than in CG, but it can also occur in CG, and, actually, even small CG examples can be constructed to illustrate it; see [29].

These investigations are easily adapted to other methods, including, *e.g.*, (Bi)CGS, where also Theorem 3 applies. Consequently, for (Bi)CGS the gap is not as bad as one might expect from the very erratic convergence behavior, because the local errors (which may be large due to high peaks in the residual norm history) are not amplified by large factors. However, the peaks may be so high that neither the updated nor the true residual converge.

A fairly general remedy against the growth of the gap between true and updated residuals — and thus against the corresponding loss of attainable accuracy — is based on an idea of Neumaier [35]: *occasional synchronization of true and updated residual combined with a shift of origin.* Neumaier's proposal is just a variation of using true instead of recursive residuals. He suggested to compute in (Bɪ)CGS the true residual at every step where the residual norm is reduced and, at the same time, to replace the current system by one for the remaining correction $\delta \mathbf{x}$ in $\mathbf{x}$, so that the current residual becomes the new right-hand side. One can think of this as a repeated shift of the origin or an *implicit iterative refinement.* At the beginning we let

$$\mathbf{b}' := \mathbf{b} - \mathbf{A}\mathbf{x}_0 , \qquad \mathbf{x}' := \mathbf{x}_0 , \qquad \mathbf{x}_0 := \mathbf{o} ,$$

so that $\mathbf{b} - \mathbf{A}\mathbf{x} = \mathbf{b}' - \mathbf{A}\,\delta\mathbf{x}$, where $\delta\mathbf{x} := \mathbf{x} - \mathbf{x}'$. We then apply our algorithm of choice to $\mathbf{A}\,\delta\mathbf{x} = \mathbf{b}'$. At step n, if the *update condition*

$$||\mathbf{r}_n|| < ||\mathbf{b}'||\,\gamma' \qquad (\text{where} \quad \gamma' \in (0, 1] \quad \text{is given}) \tag{39}$$

is satisfied, we include the reassignments

$$\mathbf{b}' := \mathbf{b}' - \mathbf{A}\mathbf{x}_n , \qquad \mathbf{x}' := \mathbf{x}' + \mathbf{x}_n , \qquad \mathbf{x}_n := \mathbf{o} . \tag{40}$$

Note that at every step, we then have

$$\mathbf{r}_n = \mathbf{b}' - \mathbf{A}\mathbf{x}_n = \mathbf{b} - \mathbf{A}(\mathbf{x}' + \mathbf{x}_n) .$$

Neumaier actually computed the true residual at every step and chose $\gamma' = 1$, which means that the update is performed at every step where the residual decreases, hence, nearly always. Sleijpen and van der Vorst [41] followed up on this idea and suggested several alternatives to the update condition (39), so that fewer shifts and true residuals are used. Recently, van der Vorst and Ye [46] came up with yet another improvement of this strategy. It pushes the level of stagnation of the true residual down to the level that has to be expected in view of the roundoff bounds for the evaluation of the residual at the rounded exact solution in finite precision arithmetic. In general, each update (40) requires an extra matrix-vector product. However, Neumaier [35] found a way to use it in (Bɪ)CGS for replacing one of the two other such products, and Sleijpen and van der Vorst [41] achieved the same for BɪCG-STAB.

Acknowledgment. The author would like to express sincere thanks to Andreas Frommer and Miroslav Rozložník for their helpful comments.

References

1. A. Boriçi. *Krylov Subspace Methods in Lattice QCD.* Diss. ETH, Swiss Federal Institute of Technology (ETH) Zurich, 1996.
2. A. Boriçi and P. de Forcrand. Fast Krylov space methods for calculation of quark propagators. IPS Research Report 94-03, ETH Zurich, 1994.
3. G. Cella, A. Hoferichter, V. K. Mitrjushkin, M. Müller-Preuss, and A. Vincere. Efficiency of different matrix inversion methods applied to Wilson fermions. Technical Report HU Berlin–EP-96/17, IFUP–TH 29/96, SWAT/96/108, 1996.
4. J. K. Cullum and R. A. Willoughby. *Lanczos Algorithms for Large Symmetric Eigenvalue Computations (2 Vols.).* Birkhäuser, Boston-Basel-Stuttgart, 1985.
5. D. Day. An efficient implementation of the non-symmetric Lanczos algorithm. *SIAM J. Matrix Anal. Appl.*, 18:566–589, 1997.
6. D. M. Day III. *Semi-duality in the two-sided Lanczos algorithm.* PhD thesis, University of California at Berkeley, 1993.
7. P. Fiebach, R. W. Freund, and A. Frommer. Variants of the block-QMR method and applications in quantum chromodynamics. In A. Sydow, editor, *15th IMACS World Congress on Scientific Computation, Modelling and Applied Mathematics, Vol. 3, Computational Physics, Chemistry and Biology*, pages 491–496. Wissenschaft und Technik Verlag, 1997.
8. R. Fletcher. Conjugate gradient methods for indefinite systems. In G. A. Watson, editor, *Numerical Analysis, Dundee, 1975*, volume 506 of *Lecture Notes in Mathematics*, pages 73–89. Springer, Berlin, 1976.
9. R. W. Freund. Conjugate gradient-type methods for linear systems with complex symmetric coefficient matrices. *SIAM J. Sci. Statist. Comput.*, 13:425–448, 1992.
10. R. W. Freund. Lanczos-type algorithms for structured non-Hermitian eigenvalue problems. In J. D. Brown, M. T. Chu, D. C. Ellison, and R. J. Plemmons, editors, *Proceedings of the Cornelius Lanczos International Centenary Conference*, pages 243–245. SIAM, Philadelphia, PA, 1994.
11. R. W. Freund, M. H. Gutknecht, and N. M. Nachtigal. An implementation of the look-ahead Lanczos algorithm for non-Hermitian matrices. *SIAM J. Sci. Comput.*, 14:137–158, 1993.
12. R. W. Freund and N. M. Nachtigal. An implementation of the QMR method based on coupled two-term recurrences. *SIAM J. Sci. Comput.*, 15:313–337, 1994.
13. R. W. Freund and N. M. Nachtigal. Software for simplified Lanczos and QMR algorithms. *Applied Numerical Mathematics*, 19:319–341, 1995.
14. A. Frommer. Linear system solvers — recent developments and implications for lattice computations. *Nuclear Physics B (Proc. Suppl.)*, 53:120–126, 1996.
15. A. Frommer, V. Hannemann, B. Nöckel, T. Lippert, and K. Schilling. Accelerating Wilson fermion matrix inversions by means of the stabilized biconjugate gradient algorithm. *Int. J. Modern Physics C*, 5:1073–1088, 1994.
16. A. Frommer and B. Medeke. Exploiting structure in Krylov subspace methods for the Wilson fermion matrix. In A. Sydow, editor, *15th IMACS World Congress on Scientific Computation, Modelling and Applied Mathematics, Vol. 3, Computational Physics, Chemistry and Biology*, pages 485–490. Wissenschaft und Technik Verlag, 1997.

17. A. Frommer, B. Nöckel, S. Güsken, T. Lippert, and K. Schilling. Many masses on one stroke: economic computation of quark propagators. *Int. J. Modern Physics C*, 6:627–638, 1995.

18. A. Greenbaum. Predicting the behavior of finite precision Lanczos and conjugate gradient computations. *Linear Algebra Appl.*, 113:7–63, 1989.

19. A. Greenbaum. Accuracy of computed solutions from conjugate-gradient-like methods. In M. Natori and T. Nodera, editors, *Advances in Numerical Methods for Large Sparse Sets of Linear Systems*, number 10 in Parallel Processing for Scientific Computing, pages 126–138. Keio University, Yokahama, Japan, 1994.

20. A. Greenbaum. Estimating the attainable accuracy of recursively computed residual methods. *SIAM J. Matrix Anal. Appl.*, 18(3):535–551, 1997.

21. A. Greenbaum and Z. Strakoš. Predicting the behavior of finite precision Lanczos and conjugate gradient computations. *SIAM J. Matrix Anal. Appl.*, 13(1):121–137, 1992.

22. M. H. Gutknecht. Solving Theodorsen's integral equation for conformal maps with the fast fourier transform and various nonlinear iterative method. *Numer. Math.*, 36:405–429, 1981.

23. M. H. Gutknecht. The unsymmetric Lanczos algorithms and their relations to Padé approximation, continued fractions, and the qd algorithm. in Preliminary Proceedings of the Copper Mountain Conference on Iterative Methods, April 1990. http://www.sam.math.ethz.ch/~mhg/pub/CopperMtn90.ps.Z and CopperMtn90-7.ps.Z.

24. M. H. Gutknecht. A completed theory of the unsymmetric Lanczos process and related algorithms, Part I. *SIAM J. Matrix Anal. Appl.*, 13:594–639, 1992.

25. M. H. Gutknecht. Variants of BiCGStab for matrices with complex spectrum. *SIAM J. Sci. Comput.*, 14:1020–1033, 1993.

26. M. H. Gutknecht. A completed theory of the unsymmetric Lanczos process and related algorithms, Part II. *SIAM J. Matrix Anal. Appl.*, 15:15–58, 1994.

27. M. H. Gutknecht. Lanczos-type solvers for nonsymmetric linear systems of equations. *Acta Numerica*, 6:271–397, 1997.

28. M. H. Gutknecht and K. J. Ressel. Look-ahead procedures for Lanczos-type product methods based on three-term recurrences. Tech. Report TR-96-19, Swiss Center for Scientific Computing, June 1996.

29. M. H. Gutknecht and Z. Strakoš. Accuracy of two three-term and three two-term recurrences for Krylov space solvers. *SIAM J. Matrix Anal. Appl.* To appear.

30. M. R. Hestenes and E. Stiefel. Methods of conjugate gradients for solving linear systems. *J. Res. Nat. Bureau Standards*, 49:409–435, 1952.

31. M. Hochbruck. The Padé table and its relation to certain numerical algorithms. Habilitationsschrift, Universität Tübingen, Germany, 1996.

32. R. A. Horn and C. R. Johnson. *Matrix Analysis*. Cambridge University Press, New York, 1985.

33. C. Lanczos. Solution of systems of linear equations by minimized iterations. *J. Res. Nat. Bureau Standards*, 49:33–53, 1952.

34. T. A. Manteuffel. The Tchebyshev iteration for nonsymmetric linear systems. *Numer. Math.*, 28:307–327, 1977.

35. A. Neumaier. Iterative regularization for large-scale ill-conditioned linear systems. Talk at Oberwolfach, April 1994.

36. B. N. Parlett. *The Symmetric Eigenvalue Problem*. Prentice-Hall, Englewood Cliffs, N.J., 1980.

37. B. N. Parlett and D. S. Scott. The Lanczos algorithm with selective reorthogonalization. *Math. Comp.*, 33:217–238, 1979.

38. B. N. Parlett, D. R. Taylor, and Z. A. Liu. A look-ahead Lanczos algorithm for unsymmetric matrices. *Math. Comp.*, 44:105–124, 1985.

39. H. Rutishauser. Beiträge zur Kenntnis des Biorthogonalisierungs-Algorithmus von Lanczos. *Z. Angew. Math. Phys.*, 4:35–56, 1953.

40. Y. Saad. The Lanczos biorthogonalization algorithm and other oblique projection methods for solving large unsymmetric systems. *SIAM J. Numer. Anal.*, 2:485–506, 1982.

41. G. L. G. Sleijpen and H. A. van der Vorst. Reliable updated residuals in hybrid Bi-CG methods. *Computing*, 56:141–163, 1996.

42. G. L. G. Sleijpen, H. A. van der Vorst, and D. R. Fokkema. BiCGstab(l) and other hybrid Bi-CG methods. *Numerical Algorithms*, 7:75–109, 1994.

43. G. L. G. Sleijpen, H. A. van der Vorst, and J. Modersitzki. Effects of rounding errors in determining approximate solutions in Krylov solvers for symmetric linear systems. Preprint 1006, Department of Mathematics, Universiteit Utrecht, March 1997.

44. P. Sonneveld. CGS, a fast Lanczos-type solver for nonsymmetric linear systems. *SIAM J. Sci. Statist. Comput.*, 10:36–52, 1989.

45. H. A. van der Vorst. Bi-CGSTAB: a fast and smoothly converging variant of Bi-CG for the solution of nonsymmetric linear systems. *SIAM J. Sci. Statist. Comput.*, 13:631–644, 1992.

46. H. A. van der Vorst and Q. Ye. Residual replacement strategies for Krylov subspace iterative methods for the convergence of true residuals. Preprint, 1999.

47. R. S. Varga. *Matrix Iterative Analysis*. Prentice-Hall, Englewood Cliffs, N.J., 1962. Rev. 2nd ed., Springer-Verlag, 1999.

48. H. E. Wrigley. Accelerating the Jacobi method for solving simultaneous equations by Chebyshev extrapolation when the eigenvalues of the iteration matrix are complex. *Comput. J.*, 6:169–176, 1963.

49. S.-L. Zhang. GPBI-CG: generalized product-type methods based on Bi-CG for solving nonsymmetric linear systems. *SIAM J. Sci. Comput.*, 18(2):537–551, 1997.

50. H. Zurmühl. *Matrizen und ihre technischen Anwendungen, 4. Aufl.* Springer-Verlag, Berlin, 1994.

An Algebraic Multilevel Preconditioner for Symmetric Positive Definite and Indefinite Problems

Arnold Reusken

Institut für Geometrie und Praktische Mathematik
RWTH Aachen
Templergraben 55, D-52056 Aachen, Germany

Abstract. We present a preconditioning method for the iterative solution of large sparse systems of equations. The preconditioner is based on ideas both from ILU preconditioning and from multigrid. The resulting preconditioning technique requires the matrix only. A multilevel structure is obtained by constructing a maximal independent set of the graph of a reduced matrix. The computation of a Schur complement approximation is based on a Galerkin approach with a matrix dependent prolongation and restriction. The resulting preconditioner has a transparant modular structure similar to the algorithmic structure of a multigrid V-cycle. The method is applied to symmetric positive definite and indefinite Helmholtz problems. The multilevel preconditioner is compared with standard ILU preconditioning methods.

1 Introduction

Multigrid methods are very efficient iterative solvers for the large systems of equations resulting from discretization of partial differential equations (cf. [11,26] and the references therein). An important principle of multigrid is that a basic iterative method, which yields appropriate local corrections, is applied on a hierarchy of discretizations with different characteristic mesh sizes. This multilevel structure is of main importance for the efficiency of multigrid.

Another class of efficient iterative solvers consists of Krylov subspace methods combined with ILU preconditioning (cf. [8,20] and the references therein). These methods only need the matrix and are in general easier to implement than multigrid methods. Also the Krylov subspace methods are better suitable for a "black-box" approach. On the other hand, for discretized partial differential equations the Krylov methods with ILU preconditioning are often less efficient than multigrid methods.

In the multigrid field there have been developed methods which have a multilevel structure but require only the matrix of the linear system. These are called *algebraic* multigrid methods. Approaches towards algebraic multigrid are presented in, e.g. [7,9,19,23,25]. In all these methods one tries to mimic the multigrid principle. First one introduces a reasonable coarse "grid" space.

Then a prolongation operator is chosen and for the restriction one usually takes the adjoint of the prolongation. The operator on the coarse grid space is defined by a Galerkin approach. With these components, a standard multigrid approach (smoothing + coarse grid correction) is applied. These algebraic multigrid methods can be used in situations where a grid (hierarchy) is not available. Also these methods can be used for developing black-box solvers.

Recently there have been developed ILU type of preconditioners with a multilevel structure, cf. [5,6,16,21,22]. The multilevel structure is induced by a level wise numbering of the unknowns. In [2,3,17,18], new hybrid methods have been presented, which use ideas both from ILU (incomplete Gaussian elimination) and from multigrid.

In the present paper we reconsider the approximate cyclic reduction preconditioner of [17,18]. This method is based on the recursive application of a two-level method, as in cyclic reduction or in a multigrid V-cycle method. For the definition of a two level structure we use two important concepts: a reduced graph and a maximal independent set. The partitioning of the set of unknowns, denoted by the labels "red " and "black", yields a corresponding block-representation of the given matrix $\mathbf{A}$:

$$\mathbf{PAP}^T = \begin{bmatrix} \mathbf{A}_{bb} & \mathbf{A}_{br} \\ \mathbf{A}_{rb} & \mathbf{A}_{rr} \end{bmatrix} , \tag{1}$$

with $\mathbf{P}$ a suitable permutation matrix. The construction of the red-black partitioning is such that, under reasonable assumptions on $\mathbf{A}$, the $\mathbf{A}_{rr}$ block is guaranteed to be strongly diagonally dominant. In [18] one can find a technique for constructing a sparse approximation $\tilde{\mathbf{S}}_{bb}$ of the Schur complement $\mathbf{S}_{bb} := \mathbf{A}_{bb} - \mathbf{A}_{br}\mathbf{A}_{rr}^{-1}\mathbf{A}_{rb}$. This approximation is obtained by replacing the *block* Gaussian elimination which results in the Schur complement (cf. (4)) by a sequence of *point* Gaussian elimination steps.

In [18] the approximate cyclic reduction preconditioner is presented and analyzed in a general framework and applied to convection-diffusion and anisotropic diffusion problems. In the present paper we explain a simple variant of the cyclic reduction preconditioner which is then applied to a discretization of the Helmholtz equation $-\Delta u - \lambda u = f$ ($\lambda \geq 0$ a constant) on the unit square. In Sect. 4 we consider $\lambda = 0$ (Poisson equation) and $\lambda = 19.73$ (SPD, close to singular) and compare CG + approximate cyclic reduction preconditioning with the standard ICCG method. In Sect. 5 we consider the indefinite case ($\lambda = 100$, $\lambda = 200$) and compare the GMRES(5)+ ILU preconditioning (using droptolerances) with GMRES(5) + approximate cyclic reduction preconditioning.

2 The Cyclic Reduction Principle

We recall the classical method of cyclic reduction. This method can be used, for example, for solving a linear system with a tridiagonal matrix or with a special block tridiagonal matrix (cf. [10,13,24]). We explain the cyclic reduction principle by considering an $n \times n$ linear system with a tridiagonal matrix:

$$
\mathbf{Ax} = \mathbf{b}, \quad \mathbf{A} = \begin{bmatrix} a_1 & b_1 & & & \\ c_1 & a_2 & b_2 & & \emptyset \\ & \ddots & \ddots & \ddots & \\ \emptyset & & \ddots & \ddots & b_{n-1} \\ & & & c_{n-1} & a_n \end{bmatrix}, \quad a_i \neq 0 \quad \text{for all } i . \tag{2}
$$

Reordering the unknowns based on an obvious red-black (or "odd-even") structure results in a permuted system with a matrix of the form

$$
\mathbf{PAP}^T = \begin{bmatrix} \mathbf{A}_{bb} & \mathbf{A}_{br} \\ \mathbf{A}_{rb} & \mathbf{A}_{rr} \end{bmatrix} , \tag{3}
$$

in which $[\mathbf{A}_{bb} \ \mathbf{A}_{br}]$ represents the equations in the unknowns with a black label and $[\mathbf{A}_{rb} \ \mathbf{A}_{rr}]$ represents the equations in the unknowns with a red label. Note that, because $\mathbf{A}$ is tridiagonal, the diagonal blocks $\mathbf{A}_{bb}, \mathbf{A}_{rr}$ are diagonal matrices. Gaussian elimination in the red points results in a reduced system with dimension (approximately) $\frac{1}{2}n$. In matrix notation this corresponds to block UL-decomposition:

$$
\mathbf{PAP}^T = \begin{bmatrix} \mathbf{I} & \mathbf{A}_{br}\mathbf{A}_{rr}^{-1} \\ \emptyset & \mathbf{I} \end{bmatrix} \begin{bmatrix} \mathbf{S}_{bb} & \emptyset \\ \mathbf{A}_{rb} & \mathbf{A}_{rr} \end{bmatrix} , \quad \mathbf{S}_{bb} := \mathbf{A}_{bb} - \mathbf{A}_{br}\mathbf{A}_{rr}^{-1}\mathbf{A}_{rb} . \tag{4}
$$

The reduced system has a matrix $\mathbf{S}_{bb}$ (Schur complement) which is tridiagonal, and thus the same approach can be applied to $\mathbf{S}_{bb}$. So the basic cyclic reduction idea is to reduce significantly the dimension of the problem repeatedly until one has a relatively small problem that can be solved easily. After this *decomposition phase* a block UL-decomposition of the matrix $\mathbf{A}$ is available and the linear system in (2) can be solved using a simple backward–forward substitution process. In this process the systems with matrix $\mathbf{A}_{rr}$ are trivial to solve, because $\mathbf{A}_{rr}$ is diagonal.

In Sect. 3 we will modify this simple direct method, resulting in a *preconditioner* for sparse matrices which are not necessarily tridiagonal. For a better understanding of this preconditioner we first give a rather detailed description of a particular implementation of the cyclic reduction method for a tridiagonal matrix, which consists of a decomposition phase and a solution phase.

Decomposition phase. We assume a tridiagonal matrix $\mathbf{A} \in \mathbb{R}^{n \times n}$. *Dimbound*, with $1 < Dimbound \ll n$ is a given integer (used in D5 below). Set $i := 1$, $\mathbf{A}_1 := \mathbf{A}$, $m_0 := n$.

D1. *Red-black partitioning* . Given the tridiagonal matrix $\mathbf{A}_i$ we construct a red-black (odd-even) partitioning of the unknowns. This results in n_i vertices with label red and m_i vertices with label black. Note: $m_i + n_i = m_{i-1}$.

D2. *Determine permutation.* We determine a symmetric permutation p_i : $\{1, 2, ..., m_{i-1}\} \rightarrow \{1, 2, ..., m_{i-1}\}$ such that applying this permutation to the index set of the unknowns results in an ordering in which all unknowns with label red have index $j \in (m_i, m_{i-1}]$ and all unknowns with label black have index $j \in [1, m_i]$. Note that since we only have to permute between the sets $\{j \mid j > m_i \text{ and label}(j) = \text{black}\}$ and $\{j \mid j \leq m_i \text{ and label}(j) = \text{red}\}$, such a permutation can be fully characterized by a permutation $\hat{p}_i : \{m_i + 1, m_i + 2, ..., m_{i-1}\} \rightarrow \{1, 2, ..., m_i\}$.

D3. *Determine permuted matrix.* The symmetric matrix corresponding to the permutation p_i of D2 is denoted by $\mathbf{P}_i$. We determine $\mathbf{P}_i \mathbf{A}_i \mathbf{P}_i$. This matrix has a 2×2-block representation:

$$\mathbf{P}_i \mathbf{A}_i \mathbf{P}_i = \begin{bmatrix} \mathbf{A}_i^{bb} & \mathbf{A}_i^{br} \\ \mathbf{A}_i^{rb} & \mathbf{A}_i^{rr} \end{bmatrix} , \tag{5}$$

with $\mathbf{A}_i^{rr} \in \mathbb{R}^{n_i \times n_i}$, $\mathbf{A}_i^{bb} \in \mathbb{R}^{m_i \times m_i}$, $\mathbf{A}_i^{rb} \in \mathbb{R}^{n_i \times m_i}$, $\mathbf{A}_i^{br} \in \mathbb{R}^{m_i \times n_i}$.

D4. *Compute Schur complement.* Compute the Schur complement $\mathbf{A}_{i+1} := \mathbf{P}_i \mathbf{A}_i \mathbf{P}_i / \mathbf{A}_i^{rr} := \mathbf{A}_i^{bb} - \mathbf{A}_i^{br} (\mathbf{A}_i^{rr})^{-1} \mathbf{A}_i^{rb}$.

D5. *Store.* Save $m_i, \hat{p}_i, \mathbf{A}_i^{rr}, \mathbf{A}_i^{rb}, \mathbf{A}_i^{br}$. If $m_i < Dimbound$ then save $\mathbf{A}_{i+1}$ (stop the reduction process) else $i := i + 1$ and goto D1.

Remark 1. The algorithm used in the decomposition phase is well-defined iff the diagonal matrices $\mathbf{A}_i^{rr}$ are nonsingular. The latter property holds if the matrix $\mathbf{A}$ is symmetric positive definite or an M-matrix. This follows from the fact that the Schur complement of an SPD-matrix (M-matrix) is an SPD-matrix (M-matrix), cf. [12].

If the above decomposition process stops with $i = i_{\max}$, we obtain integers $m_1 > m_2 > ... > m_{i_{\max}}$, permutation vectors $\hat{p}_i$ ($1 \leq i \leq i_{\max}$), sparse matrices $\mathbf{A}_i^{rr}, \mathbf{A}_i^{rb}, \mathbf{A}_i^{br}$ ($1 \leq i \leq i_{\max}$) and the approximate Schur complement on the highest level $\mathbf{A}_{i_{\max}+1}$. We use the following terminology: $\hat{p}_i$ is called the *permutation* operator on level i, $\mathbf{A}_i^{rr}$ is called the *solve* operator on level i, $\mathbf{A}_i^{rb}$ is called the *collect* operator on level i, $\mathbf{A}_i^{br}$ is called the *distribute* operator on level i.

The red unknowns on all levels, together with the black unknowns on the final level induce a direct sum decomposition $\mathbb{R}^n = \mathbb{R}^{n_1} \oplus \mathbb{R}^{n_2} \oplus \ldots \oplus \mathbb{R}^{n_{i_{max}}} \oplus \mathbb{R}^{m_{i_{max}}}$. The unknowns on level i with label red are assigned the level number i, and the vertices on level i_{max} with label black are assigned level number $i_{max} + 1$. The unknowns with level number j are called the level j unknowns. Note that every unknown has a unique level number.

Solution phase. For a clear description of the solution phase we introduce permute, collect, distribute and solve operations. These operations use the corresponding operators which are available from the decomposition phase. We give a description in a pseudo-programming language.

procedure `permuteoperation`(i: int; var $\mathbf{x} \in \mathbb{R}^{m_{i-1}}$)   (* uses $\hat{p}_i$*)
    for $j := m_i + 1$ to m_{i-1} do
    if $j \neq \hat{p}_i(j)$ then interchange x_j and $x_{\hat{p}_i(j)}$;

procedure `collectoperation`(i: int; var $\mathbf{x} \in \mathbb{R}^{n_i}$; $\mathbf{g} \in \mathbb{R}^{m_i}$)   (* uses $\mathbf{A}_i^{rb}$*)
    compute $\mathbf{x} := \mathbf{x} - \mathbf{A}_i^{rb}\mathbf{g}$;

procedure `distributeoperation`(i: int; var $\mathbf{x} \in \mathbb{R}^{m_i}$; $\mathbf{g} \in \mathbb{R}^{n_i}$) (* uses $\mathbf{A}_i^{br}$*)
    compute $\mathbf{x} := \mathbf{x} - \mathbf{A}_i^{br}\mathbf{g}$;

procedure `solveoperation`(i: int; var $\mathbf{x} \in \mathbb{R}^{n_i}$)   (* uses $\mathbf{A}_i^{rr}$*)
    solve $\mathbf{A}_i^{rr}\mathbf{w} = \mathbf{x}$. $\mathbf{x} := \mathbf{w}$.

procedure `highestlevelsolve`(var $\mathbf{x} \in \mathbb{R}^{m_{i_{max}}}$)   (* uses $\mathbf{A}_{i_{max}+1}$*)
    solve $\mathbf{A}_{i_{max}+1}\mathbf{w} = \mathbf{x}$; $\mathbf{x} := \mathbf{w}$;

Using these procedures it is easy to formulate the backward and forward substitution process, i.e. the solution phase, of the approximate cyclic reduction preconditioner. On each level i ($1 \leq i \leq i_{max} + 1$) we define `ULsolve` as follows:
    procedure `ULsolve`(i: int; var $\mathbf{f} \in \mathbb{R}^{m_{i-1}}$);
        var $\mathbf{f}_{red} \in \mathbb{R}^{n_i}$;
        begin
            if $i = i_{max} + 1$ then `highestlevelsolve`($\mathbf{f}$) else
            begin
                `permuteoperation`($i, \mathbf{f}$);
                partition $\mathbf{f} = \begin{pmatrix} \mathbf{f}_b \\ \mathbf{f}_r \end{pmatrix}$ with $\mathbf{f}_r \in \mathbb{R}^{n_i}$, $\mathbf{f}_b \in \mathbb{R}^{m_i}$;
                make a copy $\mathbf{f}_{red} := \mathbf{f}_r$;
                `solveoperation`($i, \mathbf{f}_{red}$);
                `distributeoperation`($i, \mathbf{f}_b, \mathbf{f}_{red}$);
                `ULsolve`($i + 1, \mathbf{f}_b$);
                `collectoperation`($i, \mathbf{f}_r, \mathbf{f}_b$);

```
      solveoperation(i, f_r);
      permuteoperation(i, f);
   end
end;
```

The solution of $\mathbf{Ax} = \mathbf{b}$ results from the call ULsolve(1, b). The structure of ULsolve is similar to the structure of the multigrid V-cycle algorithm as presented in [11]. The distribute and collect operations correspond to the multigrid restriction and prolongation, respectively. The solve operation corresponds to the smoother in multigrid. Note, however, that in ULsolve we do not use any grid information and that every unknown is involved in the solve operations of precisely one level (as in hierarchical basis multigrid, cf. [1]).

3   Approximate Cyclic Reduction Preconditioning

In this section we introduce an approximate cyclic reduction preconditioner. For this we recall a few notions from graph theory.

A matrix $\mathbf{A} \in \mathbb{R}^{n \times n}$ induces an *ordered directed graph* $G_A(V, E)$, consisting of an ordered set of vertices $V = \{1, 2, \ldots, n\}$ and a set E of ordered pairs of vertices called *edges* . This set E consists of all pairs (i, j) for which $a_{ij} \neq 0$. A directed graph will also be called a *digraph*. If (i, j) is an element of E then i is said to be *adjacent to* j and j is said to be *adjacent from* i. Two vertices $i \neq j$ are said to be *independent* if $(i, j) \notin E$ and $(j, i) \notin E$. A subset M of V is called an independent set if every two vertices in M are independent. M is called a *maximal independent set* of vertices if M is independent but no proper superset of M in V is independent. Note that a maximal independent set is in general not unique. For a vertex $i \in V$, its *neighbourhood* $N(i)$ is defined by $N(i) = \{j \in V \mid j \neq i$ and $(i, j) \in E\}$. For $i \in V$ its *degree* , $deg(i)$, is the number of elements in the neighbourhood of i, that is, $deg(i) = |N(i)|$. A vertex i is called an *isolated* vertex if $deg(i) = 0$. Note that an isolated vertex can be adjacent *from* other vertices in V.

In the classical cyclic reduction method, as described in Sect. 2, a red-black partitioning $V = V_r \cup V_b$, $V_r \cap V_b = \{ \ \}$ of the vertex set V corresponding to a tridiagonal matrix is constructed such that V_r is a maximal independent subset of V. Since the vertices in V_r are independent, the matrix $\mathbf{A}_{rr}$ is diagonal. Moreover, the red-black partitioning yields a maximal independent set V_r for which the corresponding Schur complement $\mathbf{A}/\mathbf{A}_{rr}$ is tridiagonal and has much smaller dimension than the original matrix. For a general sparse matrix $\mathbf{A}$ it is easy to construct a partitioning $V = V_r \cup V_b$, $V_r \cap V_b = \{ \ \}$ of the vertex set of the graph $G_A(V, E)$ such that V_r is a maximal independent subset of V. Hence the same approach as in cyclic reduction, i.e. compute the Schur complement and apply the same technique recursively, can be applied. However, it is well-known that almost always one gets an unacceptable amount of fill-in in the Schur complement after only a few recursive steps. Thus this direct (!) method is not satisfactory. In the approximate cyclic reduction preconditioner we construct a partitioning $V = V_r \cup V_b$, $V_r \cap V_b = \{\ \}$ such that $\mathbf{A}_{rr}$ is *strongly diagonally dominant*. Furthermore an *approximate* Schur complement is used and systems with matrix $\mathbf{A}_{rr}$ are solved only *approximately*. The preconditioner has the following structure, which is very similar to the cyclic reduction method of Sect. 2.

Decomposition phase. We assume a sparse matrix $\mathbf{A} \in \mathrm{I\!R}^{n \times n}$. *Dimbound*, with $1 < Dimbound \ll n$ is a given integer (used in $\overline{\mathrm{D5}}$ below). Set $i := 1$, $\mathbf{A}_1 := \mathbf{A}$, $m_0 := n$.

$\overline{\mathrm{D1}}$. *Partitioning of the vertex set.* Compute a partitioning $V_i = V_i^r \cup V_i^b$, $V_i^r \cap V_i^b = \{\ \}$ of the vertex set V_i of the graph corresponding to $\mathbf{A}_i$ such that the matrix $\mathbf{A}_i^{rr}$ is strongly diagonally dominant. This results in n_i vertices with label red and m_i vertices with label black. Note: $m_i + n_i = m_{i-1}$.
$\overline{\mathrm{D2}} = \mathrm{D2}$. (as in Sect. 2)
$\overline{\mathrm{D3}} = \mathrm{D3}$.
$\overline{\mathrm{D4}}$. *Compute approximate Schur complement.* Compute a sparse approximation $\mathbf{A}_{i+1} \in \mathrm{I\!R}^{m_i \times m_i}$ of the Schur complement $\mathbf{P}_i \mathbf{A}_i \mathbf{P}_i / \mathbf{A}_i^{rr}$.
$\overline{\mathrm{D5}}$. *Store.* Save $m_i, \hat{p}_i, \mathbf{A}_i^{rr}, \mathbf{A}_i^{rb}, \mathbf{A}_i^{br}$. If $m_i < Dimbound$ or $m_i > 0.8m_{i-1}$ then save $\mathbf{A}_{i+1}$ (stop the reduction process) else $i := i + 1$ and goto D1.

The methods used in $\overline{\mathrm{D1}}$, $\overline{\mathrm{D4}}$ will be explained below. In $\overline{\mathrm{D5}}$ we introduced an additional stopping criterion to avoid a stagnation in the problem size reduction.

Solution phase. We apply the procedure ULsolve of Sect. 2 in which all procedures are *unchanged* except for the procedure solveoperation which is replaced by

procedure solveoperation(i: int; var $\mathbf{x} \in \mathrm{I\!R}^{n_i}$)
    solve $\mathbf{A}_i^{rr} \mathbf{w} = \mathbf{x}$ *approximately*, using a few iterations of a basic iterative method. The result is written in $\mathbf{x}$.

For the choice of the basic iterative method in solveoperation there are obvious possibilities: Jacobi, Gauss-Seidel, ILU. Note that the construction in the decomposition phase is such that the matrices $\mathbf{A}_i^{rr}$ are strongly diagonally dominant, hence these basic iterative methods have a high convergence rate for these systems. In this paper we use a method of Jacobi type. The solution of the system $\mathbf{A}_i^{rr} \mathbf{w} = \mathbf{x}$ is approximated by $\mathbf{w}^3$, which results from:

$$
\begin{aligned}
\mathbf{w}^1 &= (\mathbf{D}_i^{rr})^{-1}\mathbf{x} \qquad \text{(start)} \\
\mathbf{w}^2 &= \mathbf{w}^1 - (\tilde{\mathbf{D}}_i^{rr})^{-1}(\mathbf{A}_i^{rr}\mathbf{w}^1 - \mathbf{x}) \quad \text{(modified Jacobi)} \\
\mathbf{w}^3 &= \mathbf{w}^2 - (\mathbf{D}_i^{rr})^{-1}(\mathbf{A}_i^{rr}\mathbf{w}^2 - \mathbf{x}) \quad \text{(Jacobi)},
\end{aligned}
\tag{6}
$$

with $\mathbf{D}_i^{rr} = \mathrm{diag}(\mathbf{A}_i^{rr})$, $\tilde{\mathbf{D}}_i^{rr}$ the diagonal matrix which satisfies $\tilde{\mathbf{D}}_i^{rr}\mathbb{1} = \mathbf{A}_i^{rr}\mathbb{1}$, $\mathbb{1} := (1,1,\dots,1)^T$. Using $\mathbf{M} = \mathbf{I} - (\mathbf{D}_i^{rr})^{-1}\mathbf{A}_i^{rr}$, $\tilde{\mathbf{M}} = \mathbf{I} - (\tilde{\mathbf{D}}_i^{rr})^{-1}\mathbf{A}_i^{rr}$, a simple computation yields that $\mathbf{w}^3 = (\mathbf{I} - \mathbf{M}\tilde{\mathbf{M}}\mathbf{M})(\mathbf{A}_i^{rr})^{-1}\mathbf{x}$. If $\mathbf{A}_i^{rr}$ is symmetric positive definite then the matrix $(\mathbf{I} - \mathbf{M}\tilde{\mathbf{M}}\mathbf{M})(\mathbf{A}_i^{rr})^{-1}$ is symmetric. This conservation of symmetry is important for the CG method in Sect. 4. In general the rate of convergence of the Jacobi method (iteration matrix $\mathbf{M}$) is higher than the rate of convergence of the modified Jacobi method (iteration matrix $\tilde{\mathbf{M}}$). However, for the modified Jacobi method the (consistency) property $\tilde{\mathbf{M}}\mathbb{1} = 0$ holds, which is favourable for Poisson type of problems (cf. [15]). This motivates our choice for a symmetric combination of the Jacobi and the modified Jacobi method.

This procedure `solveoperation` is used both in Sect. 4 and in Sect. 5.

We now explain methods which can be used in in $\overline{\mathrm{D1}}$ and $\overline{\mathrm{D4}}$. Numerical experiments based on these methods are presented in Sect. 4 and Sect. 5.

Partitioning of the vertex set. The partitioning method consists of three steps, where the third one is optional (cf. Sect. 4 and 5).
P1. Compute a *reduced* digraph.
As in algebraic multigrid methods (cf. [19,25]), for the graph coarsening we distinguish "strong" and "weak" edges in the digraph. The underlying multigrid heuristic is that if one wants to use simple (point) smoothers then one should coarsen in the direction of the "strong" connections.
Every loop in E, i.e. an edge of the form (i,i), is labeled strong. For every nonisolated vertex $i \in V$ an edge $(i,j) \in E$ with $j \neq i$ is labeled strong if for the corresponding matrix entry a_{ij} we have:

$$|a_{ij}| \geq \beta \max_{j \in N(i)} |a_{ij}| \ , \tag{7}$$

with $0 \leq \beta < 1$ a given parameter (typically $0.4 \leq \beta \leq 0.6$). An edge is labeled weak if it is not strong. Note that for every nonisolated vertex i there is at least one strong edge (i,j) with $j \neq i$. Thus we obtain a partitioning $E = E_s \cup E_w$ of the edges into strong (E_s) and weak (E_w) edges. The directed graph consisting of the vertex set V and the set of strong edges E_s is called the *reduced* digraph and is denoted by $G_A(V, E_s)$.
P2. Compute a maximal independent set of the reduced digraph. We compute a maximal independent set M of the reduced digraph $G_A(V, E_s)$. This can be realized with low computational costs using a simple breadth first search technique (cf. [14,18]). A vertex $i \in V$ is assigned a red (black) label if $i \in M$ ($i \notin M$).
P3.(optional) Check for diagonal dominance. The vertex set partitioning constructed in P1, P2 results in a corresponding block partitioning of the matrix $\mathbf{A}$ as in (1). We now check diagonal dominance of the $\mathbf{A}_{rr}$ block. If for a

given parameter value κ

$$\sum_j |(\mathbf{A}_{rr})_{i,j}| > \kappa |(\mathbf{A}_{rr})_{i,i}| \qquad (8)$$

for some red vertex i, then the red label of this vertex is changed to black. In our applications we use $\kappa = 1.5$.

Approximate Schur complement. In [18] on can find a technique for approximating the Schur complement which is based on replacing the *block* Gaussian elimination (cf. (4)) by a sequence of *point* Gaussian elimination steps. In the present paper we use a simple variant of this technique that can be interpreted as a Galerkin approach with matrix dependent prolongation and restriction. Note that for a matrix of the form

$$\mathbf{A}^{(0)} = \begin{bmatrix} \mathbf{A}_{bb} & \mathbf{A}_{br} \\ \mathbf{A}_{rb} & \mathbf{A}_{rr} \end{bmatrix} ,$$

the Schur complement $\mathbf{S}_{bb} = \mathbf{A}^{(0)}/\mathbf{A}_{rr}$ can be represented as

$$\mathbf{S}_{bb} = \begin{bmatrix} \mathbf{I}_b & -\mathbf{A}_{br}\mathbf{A}_{rr}^{-1} \end{bmatrix} \mathbf{A}^{(0)} \begin{bmatrix} \mathbf{I}_b \\ * \end{bmatrix} = \begin{bmatrix} \mathbf{I}_b & * \end{bmatrix} \mathbf{A}^{(0)} \begin{bmatrix} \mathbf{I}_b \\ -\mathbf{A}_{rr}^{-1}\mathbf{A}_{rb} \end{bmatrix} , \qquad (9)$$

with $*$ arbitrary. Let $\mathbf{D}_{rr} = \mathrm{diag}(\mathbf{A}_{rr})$ and $\tilde{\mathbf{D}}_{rr}$ be the diagonal matrix which satisfies $\tilde{\mathbf{D}}_{rr}\mathbb{1} = \mathbf{A}_{rr}\mathbb{1}$ and let

$$\tilde{\mathbf{p}}_A = \begin{bmatrix} \mathbf{I}_b \\ -\tilde{\mathbf{D}}_{rr}^{-1}\mathbf{A}_{rb} \end{bmatrix} , \quad \tilde{\mathbf{r}}_A = \begin{bmatrix} \mathbf{I}_b & -\mathbf{A}_{br}\tilde{\mathbf{D}}_{rr}^{-1} \end{bmatrix} ,$$

$$\mathbf{r}_A = \begin{bmatrix} \mathbf{I}_b & -\mathbf{A}_{br}\mathbf{D}_{rr}^{-1} \end{bmatrix} , \quad \mathbf{r}_{\mathrm{inj}} = \begin{bmatrix} \mathbf{I}_b & \emptyset \end{bmatrix} .$$

For the approximation of the Schur complement in (9) we will use one of the following two Galerkin operators:

S1. <u>Galerkin approximation</u> $\hat{\mathbf{S}}_{bb}^{(1)} = \tilde{\mathbf{r}}_A \mathbf{A}^{(0)} \tilde{\mathbf{p}}_A, \quad \hat{\mathbf{S}}_{bb}^{(2)} = \mathbf{r}_A \mathbf{A}^{(0)} \tilde{\mathbf{p}}_A.$ (10)

The approximation $\hat{\mathbf{S}}_{bb}^{(1)}$ is spd if $\mathbf{A}^{(0)}$ is spd. ¿From the analysis in [18] it follows that the approximation $\hat{\mathbf{S}}_{bb}^{(1)}$ has better consistency properties than $\hat{\mathbf{S}}_{bb}^{(2)}$, whereas the latter has better stability properties.

To reduce the amount of fill-in in the approximate Schur complement $\hat{\mathbf{S}}_{bb}^{(k)}$ we use standard techniques: Restriction to a prescribed pattern (S2) and thresholding (S3). In the applications in Sect. 4, 5 we always use S2, whereas S3 is optional.

S2. <u>Restriction to prescribed pattern</u>

$$\tilde{\mathbf{S}}_{bb}^{(k)} := \left(\hat{\mathbf{S}}_{bb}^{(k)} \right)\Big|_{\mathrm{graph}(\mathbf{r}_{\mathrm{inj}}\mathbf{A}^{(0)}\tilde{\mathbf{p}}_A)} , \quad k = 1, 2. \qquad (11)$$

Here we use (with $\mathbf{C}, \mathbf{B} \in \mathbb{R}^{n \times n}$) the notation $(\mathbf{C}|_{\text{graph}(\mathbf{B})})_{i,j} = C_{i,j}$ if $i \neq j$ and $(i,j) \in \text{graph}(\mathbf{B})$, $(\mathbf{C}|_{\text{graph}(\mathbf{B})})_{i,j} = 0$ if $i \neq j$ and $(i,j) \notin \text{graph}(\mathbf{B})$ and $\text{diag}(\mathbf{C}|_{\text{graph}(\mathbf{B})})$ such that $\mathbf{C}|_{\text{graph}(\mathbf{B})} \mathbb{1} = \mathbf{C}\mathbb{1}$ (i.e. entries outside the pattern are added to the diagonal).

S3. Thresholding. We use a threshold parameter $\varepsilon_t > 0$. Let $\tilde{\mathbf{S}}_{bb} \in \mathbb{R}^{m \times m}$ be an approximate Schur complement. For $1 \leq i \leq m$, let $k_i > 0$ be the number of nonzero entries in the ith row of $\tilde{\mathbf{S}}_{bb}$. In row i every entry $(\tilde{\mathbf{S}}_{bb})_{i,j}$ with

$$|(\tilde{\mathbf{S}}_{bb})_{i,j}| < \varepsilon_t \frac{1}{k_i} \sum_j |(\tilde{\mathbf{S}}_{bb})_{i,j}| \tag{12}$$

is replaced by zero. This is done for all rows $i = 1, 2, \ldots, m$.

4 Application to a SPD Helmholtz Problem

In this section we show results of a few numerical experiments with the approximate cyclic reduction (CR) preconditioner. We consider the standard 5-point finite difference discretization of the Helmholtz problem $-\Delta u - \lambda u = f$ on $(0,1)^2$ with zero Dirichlet boundary conditions on a uniform grid with mesh size h. The smallest eigenvalue of the discrete operator is $\lambda_{\min} = 8h^{-2}\sin^2(\frac{1}{2}\pi h) - \lambda = 2\pi^2 - \lambda + \mathcal{O}(h^2) = 19.73921 - \lambda + \mathcal{O}(h^2)$. We take $\lambda = 0$, i.e. the Poisson equation, and $\lambda = 19.73$. In both cases the discrete problem is symmetric positive definite. We will compare the CG method with CR preconditioning (CR-CG) with the standard ICCG method. The implementation is done in MATLAB and we used the MATLAB function CHOLINC. We consider mesh sizes $h = \frac{1}{60}$ (symbol '+' in the figures), $h = \frac{1}{120}$ (symbol '*') and $h = \frac{1}{240}$ (symbol 'o'). In all experiments we take the righthand side such that the discrete solution is given by $\mathbb{1}/\|\mathbb{1}\|_2$ and we use zero as the starting vector.

Experiment 1. We take $\lambda = 0$. We choose the parameter value $Dimbound = 50$ in the decomposition phase. In the CR-preconditioner we use in step $\overline{\text{D1}}$ in the decomposition phase the method P1, P2 described in Sect. 3. We take $\beta = 0.6$ in (7). For this problem we do not need the check for diagonal dominance in P3 (this follows from the analysis in [18] and is confirmed by numerical experiments). In step $\overline{\text{D4}}$ in the decomposition phase we use the method as in S1, S2 (i.e. (10), (11)) with $k = 1$, i.e. the symmetric variant. We do not use the thresholding strategy described in S3. Numerical results for the ICCG and CR-CG methods are shown in Fig. 1 and Fig. 2. In Fig. 1 one can observe the well-known dependence of the convergence rate of the ICCG method on the mesh size h. When h is halved then, in order to obtain a fixed error reduction, one needs approximately twice as many ICCG iterations. This h dependence is much weaker for the CR-CG method (although there still seems to be a mild h-dependence). Also note that the ICCG method shows relatively slow convergence in the first phase of the solution process (superlinear convergence behaviour), whereas the CR-CG method has an almost linear convergence behaviour. For $h = \frac{1}{240}$, to reduce the starting error with a factor 100 one needs about 70 ICCG iterations but only 2 CR-CG iterations.

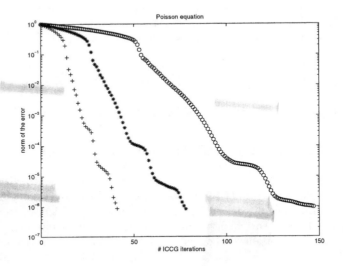

Fig. 1. ICCG.

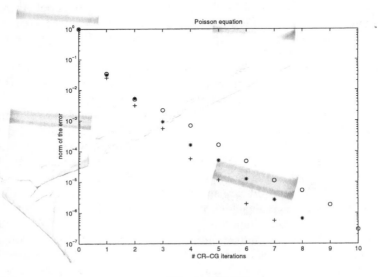

Fig. 2. CR-CG.

Remark 2. We give an indication of the storage needed for the methods in Experiment 1 for the case $h = \frac{1}{240}$, i.e. $n = 239^2$. For the symmetric matrix $\mathbf{A}$ storage for approximately $2\frac{1}{2}n$ entries is needed. The blocks $\mathbf{A}_1^{rr}$, $\mathbf{A}_1^{br}$, $\mathbf{A}_1^{rb}$, constructed in the first step (i.e. $i = 1$) of the decomposition phase of the CR-preconditioner, are blocks from the matrix $\mathbf{A}$, after a suitable permutation. Hence one does not need additional storage for these blocks. The union of the blocks $\mathbf{A}_i^{br} = (\mathbf{A}_i^{rb})^T$ over all levels $i > 1$ contains approximately $2.4n$ nonzero entries. For the union of the symmetric blocks $\mathbf{A}_i^{rr}$ over all levels $i > 1$ we need storage for approximately $1.3n$ entries. The storage needed for the approximate Schur complement on the highest level is negligible. It follows that for the CR-preconditioner (additional) storage for approximately $3.7n$ entries is needed. For the incomplete Cholesky preconditioner storage for approximately $3n$ entries is needed. Hence, for this problem, the CR-preconditioner needs only 20–30 percent more storage than the IC-preconditioner. The numerical experiments show that this statement also holds for the cases $h = \frac{1}{60}$ and $h = \frac{1}{120}$ in Experiment 1.

We briefly discuss the arithmetic work, for $h = \frac{1}{240}$, needed for one evaluation of the CR preconditioner (i.e. one call of `ULsolve(1,b)`). As a unit of arithmetic work we use one matrix-vector multiplication with the given matrix $\mathbf{A}$, denoted by MATVEC. The total arithmetic work needed in the `distributeoperation` and `collectoperation` over all levels $i \geq 1$ is approximately 2 MATVEC. In the two calls of `solveoperation` on level i we need arithmetic work comparable to 4 Jacobi iterations applied to a system with matrix $\mathbf{A}_i^{rr}$. Adding these costs over all levels $i \geq 1$ results in approximately $2\frac{1}{2}$ MATVEC arithmetic work. The costs for `highestlevelsolve` are negligible. It follows that the arithmetic costs for one CR-CG iteration are approximately $2\frac{1}{2}$ times the costs of one ICCG iteration. The same statement holds for the cases $h = \frac{1}{60}$ and $h = \frac{1}{120}$ in Experiment 1.

Experiment 2. We consider $\lambda = 19.73$. Application of the MATLAB function CHOLINC yields a well-defined incomplete Cholesky factorization of the matrix $\mathbf{A}$. The results for the ICCG method are shown in Fig. 3. We take all components and all parameter values in the CR algorithm as in Experiment 1. The results for the CR-CG method are given in Fig. 4. It turns out that, both with respect to storage and with respect to computational costs per iteration, results very similar to those formulated in Remark 2 hold. Note that the CR-CG algorithm is much more efficient than the ICCG method, but both methods have a large stagnation phase at the beginning.

5   Application to an Indefinite Helmholtz Problem

In this section we consider a discrete Helmholtz problem as in Sect. 4 (with $h = \frac{1}{60}, \frac{1}{120}, \frac{1}{240}$) but now for $\lambda = 100$, $\lambda = 200$. In these cases the problem is indefinite. In all three cases, $h = \frac{1}{60}, \frac{1}{120}, \frac{1}{240}$, the discrete operator has 6 negative eigenvalues (counted with multiplicity) if $\lambda = 100$ and 13 negative

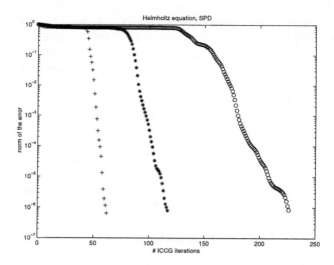

Fig. 3. ICCG.

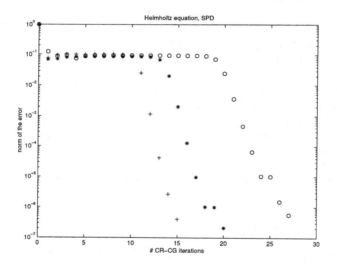

Fig. 4. CR-CG.

eigenvalues if $\lambda = 200$. The CG method is no longer applicable. The MINRES method could be used for this type of problem. If, however, one wants to combine this method with preconditioning then one needs a symmetric positive definite preconditioner. It is not clear how to construct an efficient spd preconditioner for this problem. We will use a (preconditioned) GMRES method with restart after 5 iterations (GMRES(5)) as a solver. For the preconditioner we use standard ILU techniques and the approximate CR method. For the

ILU preconditioner the MATLAB function LUINC is applied which can be used for computing the standard ILU(0) factorization and for computing an ILU factorization based on droptolerances (ILU(eps), where eps denotes the drop tolerance). The righthand side and starting vector are as in Sect. 4.

Experiment 3. We take $\lambda = 100$. The ILU(0) and ILU(eps) factorizations are computed using the MATLAB function LUINC. In Table 5 we give the number of nonzero entries in the preconditioner (where we do not make use of symmetry). The convergence behaviour of the GMRES(5) method with ILU left preconditioning (ILU-GMRES(5)) is shown for $h = \frac{1}{60}$, $h = \frac{1}{120}$ in Fig. 5 and Fig. 6, respectively. In these figures we use the following symbols: '+' for ILU(0), 'x' for ILU(0.01), 'o' for ILU(0.005), '*' for ILU(0.002). Note that the unit on the horizontal axis is one preconditioned GMRES(5) iteration, which consists of 5 preconditioned GMRES iterations. In Fig. 5 and Fig. 6 we see slow convergence and stagnation phases. Moreover, the dependence of the preconditioner on the threshold parameter eps is unpredictable. For example, for $h = \frac{1}{120}$ the result for $eps = 0.002$ is significantly better than for $eps = 0.005$ (after 100 GMRES(5) iterations), whereas for $h = \frac{1}{60}$ it is the other way round.

Table 1. Number of nonzero entries in ILU preconditioner

	ILU(0)	ILU(0.01)	ILU(0.005)	ILU(0.002)
$h = \frac{1}{60}$	17169	36828	54833	86659
$h = \frac{1}{120}$	70329	152583	218851	324555

Numerical experiments for the case with $\lambda = 200$ show a similar unsatisfactory behaviour of GMRES(5) with ILU preconditioning.

Experiment 4. We consider the indefinite problem with $\lambda = 100$, $\lambda = 200$ and apply GMRES(5) with CR preconditioning. If in the decomposition phase we use the same components and parameter values as in Experiments 1 and 2 then the resulting preconditioner is not satisfactory. The main cause for this poor behaviour lies in the fact that if we only use the method P1, P2 in the partitioning step $\overline{D1}$ then for indefinite problems (strong) diagonal dominance of the $\mathbf{A}_{rr}$ is not guaranteeed. Hence, in addition to P1, P2 we now also use the optional method in P3. This yields a significant improvement of the performance of the preconditioner. In view of the better stability properties, we take the Schur complement approximation $\hat{\mathbf{S}}_{bb}^{(2)}$ in S1, (10). A significant further improvement can be obtained if we allow more fill-in in the preconditioner. For this we simply lower the value of the parameter β in (7). This causes slower coarsening and more fill-in. Based on numerical experiments for an indefinite problem ($\lambda = 200$) of relatively low dimension ($h = \frac{1}{40}$) we choose the parameter value $\beta = 0.4$. The increase of fill-in, due this choice

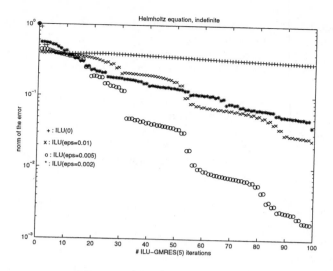

Fig. 5. ILU-GMRES(5), $h = \frac{1}{60}$.

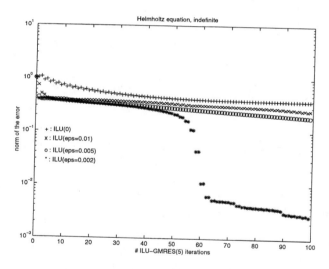

Fig. 6. ILU-GMRES(5), $h = \frac{1}{120}$.

of β, can become very large if one goes to higher levels. However, many fill-in entries turn out to be very small. Hence we now also use the optional thresholding step S3. Based on numerical experiments for a low dimensional problem we take the threshold parameter $\varepsilon_t = 0.001$ in (12). Summarizing, in the decomposition phase we use P1, P2, P3 (in $\overline{D1}$), S1 (with $\hat{S}_{bb}^{(2)}$), S2, S3 (in $\overline{D4}$) with parameter values $Dimbound = 50$, $\beta = 0.4$, $\varepsilon_t = 0.001$. The results of the GMRES(5) method with CR preconditioner are shown in Fig. 7 ($\lambda = 100$) and Fig. 8 ($\lambda = 200$). The use of the symbols '+','*','o' is as in Sect. 4.

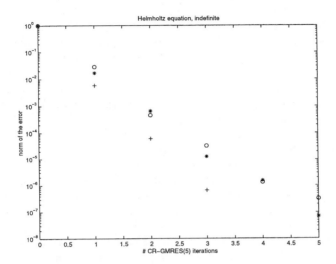

Fig. 7. CR-GMRES(5), $\lambda = 100$.

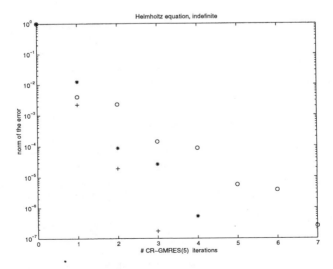

Fig. 8. CR-GMRES(5), $\lambda = 200$.

We give an indication of the storage and arithmetic work needed for the CR preconditioner (cf. Remark 2). We consider the case $h = \frac{1}{240}, \lambda = 200$, i.e. $n = 239^2$. Due to the use of the nonsymmetric Schur complement preconditioner $\hat{\mathbf{S}}_{bb}^{(2)}$ in (10) the preconditioner is in general not symmetric. The union of the blocks $\mathbf{A}_i^{br}$, $\mathbf{A}_i^{rb}$ over all levels $i > 1$ contains approximately $25n$ nonzeros entries. The union of the blocks $\mathbf{A}_i^{rr}$ over all levels $i > 1$ contains approximately $5n$ nonzeros entries. It follows that for this preconditioner the storage needed is approximately 6 times as high as for the matrix $\mathbf{A}$ (if we do not make use of symmetry). Hence these storage costs are quite high (cf. Remark 3).

The total arithmetic work needed in the `collectoperation` and in the `distributeoperation` over all levels $i \geq 1$ is approximately $5\frac{1}{2}$ MATVEC. In the two calls of `solveoperation` over all levels $i \geq 1$ we need arithmetic work comparable to 5 MATVEC.

Similar results, both with respect to storage and with respect to arithmetic work , hold for the other cases ($h = \frac{1}{60}, \frac{1}{120}, \lambda = 100$). For this type of indefinite problem the storage and arithmetic costs appear to be high. Note, however, that these problems are known to be very hard for other iterative solvers like geometric multigrid and Krylov methods with ILU preconditioning (cf. Experiment 3). From Fig. 8 we see that, for the case $h = \frac{1}{240}$, after 7 CR-GMRES(5) iterations the error has been reduced with a factor 10^7, i.e. a factor 10 per CR-GMRES(5) iteration, which corresponds to a factor 1.6 per preconditioned GMRES iteration.

Remark 3. Note that, opposite to ILU preconditioners, the solution phase of the CR preconditioner is easy to parallelize. This parallelization can be realized along the same lines as for a geometric multigrid V-cycle (cf. [4]).

References

1. Bank, R.E., Dupont, T.F., Yserentant, H.: The hierarchical basis multigrid method. Numer. Math. **52** (1988) 427–458
2. Bank, R.E., Smith, R.K.: The incomplete factorization multigraph algorithm. SIAM J. Sci. Comput. **20** (1999) 1349–1364
3. Bank, R.E., Wagner, C.: Multilevel ILU decomposition. Numer. Math **82** (1999) 543–576
4. Bastian, P.: Parallele adaptive Mehrgitterverfahren. Teubner Skripten zur Numerik, Teubner, Stuttgart, Leipzig (1996)
5. Botta, E.E.F., Van der Ploeg, A.: Preconditioning techniques for matrices with arbitrary sparsity patterns. In: Proceedings of the Ninth International Conference on Finite Elements in Fluids, New Trends and Applications (1995) 989–998
6. Botta, E.E.F., Wubs, W.: MRILU: it's the preconditioning that counts. Report W-9703, Department of Mathematics, University of Groningen, The Netherlands (1997)
7. Braess, D.: Towards algebraic multigrid for elliptic problems of second order. Computing **55** (1995) 379–393

8. Bruaset, A.M.: A survey of preconditioned iterative methods. Pitman Research Notes in Mathematics **328** Longman (1995)
9. Dendy, J.E.: Black box multigrid. J. Comput. Phys. **48** (1982) 366–386
10. Golub, G.H., Van Loan, C.: Matrix Computations. Johns Hopkins University Press, second edition (1989)
11. Hackbusch, W.: Multigrid methods and applications. Springer, Berlin, Heidelberg, New York (1985)
12. Hackbusch, W.: Iterative solution of large sparse systems of equations. Springer, New York (1994)
13. Heller, D.: Some aspects of the cyclic reduction algorithm for block tridiagonal linear systems. SIAM J. Numer. Anal. **13** (1976) 484–496
14. Horowitz, E., Sahni, S.: Fundamentals of data structures in Pascal. Pitman, London (1984)
15. Notay, Y.: Using approximate inverses in algebraic multilevel preconditioning. Numer. Math. **80** (1998) 397–417
16. Van der Ploeg, A.: Preconditioning for sparse matrices with applications. PhD thesis, University of Groningen (1994)
17. Reusken, A.: Approximate cyclic reduction preconditioning. In: Multigrid methods 5 (W. Hackbusch and G. Wittum, eds.). Lecture Notes in Computational Science and Engineering **3** (1998) 243–259
18. Reusken, A.: On the approximate cyclic reduction preconditioner. SIAM J. Sci. Comput., to appear
19. Ruge, J.W., Stüben, K.: Algebraic multigrid. In: Multigrid Methods (S.F. McCormick, ed.). SIAM, Philadelphia (1987) 73–130
20. Saad, Y.: Iterative methods for sparse linear systems. PWS Publishing Company, Boston (1996)
21. Saad, Y.: ILUM: a multi-elimination ILU preconditioner for general sparse matrices. SIAM J. Sci. Comput. **17** (1996) 830–847
22. Saad, Y., Zhang, J.: BILUM: block versions of multi-elimination and multilevel ILU preconditioner for general sparse linear systems. Report UMSI 97/126, Department of Computer Science, University of Minnesota (1997)
23. Stüben, K.: Algebraic multigrid (AMG): An introduction with applications. GMD Report **53** (1999)
24. Swarztrauber, P.N.: The methods of cyclic reduction, Fourier analysis and the FACR algorithm for the discrete solution of Poisson's equation on a rectangle. SIAM Review **19** (1977) 490–501
25. Vanek, P., Mandel, J., Brezina, M.: Algebraic multigrid by smoothed aggregation for second and fourth order elliptic problems. Computing **56** (1996) 179–196
26. Wesseling, P.: An introduction to multigrid methods. Wiley, Chichester (1992)

On Algebraic Multilevel Preconditioning

Yvan Notay*

Service de Métrologie Nucléaire
Université Libre de Bruxelles (C.P. 165–84)
50, Av. F.D. Roosevelt, B-1050 Brussels, Belgium

Abstract. Algebraic multilevel preconditioners are based on a block incomplete factorization process applied to the system matrix partitioned in hierarchical form. They have as key ingredient a technique to derive a (sparse) approximation to the Schur complement resulting from the elimination of the fine grid nodes. Once such a relevant approximation is found, it is relatively easy to set up efficient two- and multi-level schemes. Compared with more standard (classical or algebraic) multi-grid methods, an obvious advantage of this approach is that it does not require smoothing, i.e. the convergence properties do not critically depend on the good interaction between a smoother and the coarse grid correction. As a consequence, the analysis can be done purely algebraically, independently of an underlying PDE, and the technique is potentially more easily extended to other grid based computations than those directly connected to discrete elliptic PDEs.

In this chapter, we review some recent results on these methods, concerning both self-adjoint and non self-adjoint discrete second order PDEs.

1   Introduction

We consider the iterative solution of large sparse $n \times n$ linear systems

$$A\,\mathbf{u} \;=\; \mathbf{b}\,. \tag{1}$$

with real coefficient matrix A. As is well known, modern techniques combine the use of a Krylov subspace method with a proper preconditioning (e.g. [3,17,27]). A Krylov subspace method, or "Krylov accelerator", is a device to accelerate the convergence of basic iterations of the form

$$\mathbf{u}^{(k+1)} \;=\; \mathbf{u}^{(k)} + B^{-1}\left(\mathbf{b} - A\,\mathbf{u}^{(k)}\right)\,, \tag{2}$$

where B is the preconditioner. This acceleration is obtained by allowing $\mathbf{u}^{(k+1)}$ to depend also on former iterates, and by introducing parameters computed in function of some inner products. We refer to the contribution of M. Gutknecht in this volume for more details, noting only that the conjugate gradient method is mostly used when A and B are both symmetric and positive definite (the "symmetric case" in the remaining of this chapter).

* Supported by the "Fonds National de la Recherche Scientifique", Maître de recherches.

The focus of this contribution is on the design of efficient preconditioners. A preconditioner is a matrix B which is as close as possible to the system matrix A, but such that solving a system $B\mathbf{x} = \mathbf{y}$ is relatively cheap. Indeed, such a system has to be solved once per iteration, whereas the convergence rate depends on how well B approximates A (The iteration (2) converges in one step if $B = A$). More precisely, the convergence depends mainly of the eigenvalues of $B^{-1}A$, that are to be as far as possible well clustered away from the origin of the complex plane. For instance, in the symmetric case, the eigenvalues of $B^{-1}A$ are real and the number of iterations is generally proportional to the square root of

$$\kappa(B^{-1}A) = \frac{\lambda_{\max}(B^{-1}A)}{\lambda_{\min}(B^{-1}A)},$$

referred to as the *condition number* of the preconditioned system [1].

Here, we more particularly discuss so-called algebraic multilevel preconditioning methods [2,6,7,9–12,19–22,24–26,30–34], which are based on a block incomplete factorization process applied to the system matrix partitioned in hierarchical form. Note that this hierarchy may be computed from A in a black box fashion as in so-called algebraic multigrid (AMG) methods (e.g. [29] and the references therein). However algebraic *multilevel* methods have so far been essentially considered on regular grids in combination with standard geometric coarsening (say, $h-2\,h$). Besides the fact that this facilitates theoretical developments, a reason might be a better robustness. Indeed, standard multigrid methods based on interpolation between different grids [18] do not work well with geometric coarsening in some circumstances, e.g. in presence of anisotropy, whereas most algebraic multilevel schemes do not present such weaknesses. Hence, with the latter, there is no serious reason to leave out the advantages of geometric coarsening whenever working with matrices defined on regular grids. Indeed, the regularity of the grid is then preserved throughout all successive levels, which not only facilitates the analysis, but also makes the methods easier to implement and the corresponding code potentially much faster, indirect addressing schemes being not needed.

Now, before entering the details, we have to mention that all results available so far on algebraic multilevel methods were obtained in the context of systems arising from the discretization of PDEs of the form

$$-\partial_x\, a_x\, \partial_x\, u - \partial_y\, a_y\, \partial_y\, u\ (-\partial_z\, a_z\, \partial_z\, u\,) + \overline{v}\cdot\overline{\nabla} u = f\,, \qquad (3)$$

where $\overline{v}$ is a (given) convective flow. Thus, when $\overline{v} = 0$, the problem is self-adjoint and A is symmetric and positive definite, whereas A is anyway positive real ($\mathbf{v}^T A\,\mathbf{v} \geq 0\ \ \forall \mathbf{v} \in \mathcal{R}^n$) in the general case.

---

[1] Throughout this chapter, when both G and H are symmetric and positive definite (i.e. the eigenvalues of $G^{-1}H$ are real), $\lambda_{\max}\left(G^{-1}H\right)$ and $\lambda_{\min}\left(G^{-1}H\right)$ denote respectively the largest and the smallest eigenvalue of $G^{-1}H$.

In this chapter, we essentially review our recent results from [19–22], whose focus is on finite difference matrices, whereas other works in the field mostly consider finite element matrices with hierarchical basis functions. As stable finite difference schemes for the equation (3) result in a system (1) whose coefficient matrix A is a M–matrix with nonnegative row-sum, we shall in the following freely use the assumption that A belongs to this class of matrices. For the sake of clarity, let us then recall that the latter may be equivalently defined as the set of matrices with nonpositive offdiagonal entries that are (weakly or strictly) diagonally dominant. Thus, these matrices may be arbitrarily ill conditioned (may have eigenvalues very close to the origin), but satisfy nevertheless some nice algebraic properties that facilitate the development of preconditioning techniques. In this respect, it also worths noting that, when A has nonnegative row-sum and is positive real but is not an M–matrix, it is possible to first compute a nearby M–matrix using the techniques in [3,5,28], and next compute the preconditioner from the latter M–matrix. At least in the symmetric case, this approach has been proved efficient in several cases, see [5,28] for examples.

Note finally that if A is diagonally dominant but not an M–matrix, then the picture is even better because the presence of positive offdiagonal entries implies that the row-sum is strongly positive, and hence the eigenvalues are likely to be away from the origin. Therefore, the original conditioning of A is not that bad, and more standard preconditioning techniques (that are well defined for diagonally dominant matrices [3]) should be efficient enough.

2   The "Ideal" Two-Level Preconditioner

Algebraic two-level preconditioners are based on a partitioning of the unknowns in fine and coarse grid ones. Let

$$A = \begin{pmatrix} A_{11} & A_{12} \\ A_{21} & A_{22} \end{pmatrix} , \tag{4}$$

be the corresponding 2×2 block form of the system matrix, where the first block of unknowns correspond to the fine grid nodes and the second block of unknowns to the coarse grid nodes. An exact block factorization of A writes

$$A = \begin{pmatrix} A_{11} & \\ A_{21} & S_A \end{pmatrix} \begin{pmatrix} I & A_{11}^{-1}A_{12} \\ & I \end{pmatrix}$$

where

$$S_A = A_{22} - A_{21} A_{11}^{-1} A_{12}$$

is the Schur complement. In general, A_{11}^{-1} is dense and is not computable at a reasonable cost. Algebraic multilevel methods have therefore as key ingredient a technique to derive a (sparse) approximation S to S_A. If systems of the type $A_{11} \mathbf{x}_1 = \mathbf{y}_1$ can be solved exactly, the preconditioner then writes

$$B = \begin{pmatrix} A_{11} & \\ A_{21} & S \end{pmatrix} \begin{pmatrix} I & A_{11}^{-1} A_{12} \\ & I \end{pmatrix} . \tag{5}$$

A_{11} has usually most of its rows strongly diagonally dominant and for this reason is fairly well conditioned, independently of the problem (mesh) size [1,8,14]. Hence, systems with A_{11} can in principle be solved relatively cheaply by classical iterative methods. Although in general, for cost efficiency reasons, only approximate solves are performed, we may thus consider for a while the "ideal" preconditioner (5). Issues related to the approximate inversion of A_{11} are discussed in the next section.

Now, the theoretical foundation of algebraic two-level methods lies in the identity [31]

$$B^{-1} A = \begin{pmatrix} I & * \\ & S^{-1} S_A \end{pmatrix} , \tag{6}$$

so that the quality of the preconditioner essentially depends on a good clustering of the eigenvalues of $S^{-1} S_A$.

In the finite element context, most methods proposed so far take S equal to the coarse grid discretization matrix, which is naturally available through a basis transformation (In this way, one generalizes the hierarchical basis multigrid method [13–15,35]). In the finite difference context, this would require re-discretization, so we propose in [21,22] a more algebraic alternative. Letting Δ be the diagonal matrix with same row-sum as A_{11}, we take S equal to the Schur complement of

$$\widetilde{A} = \begin{pmatrix} \Delta & A_{12} \\ A_{21} & A_{22} \end{pmatrix} ,$$

in which one has deleted the zero lines and columns. In other words,

$$S = A_{22} - A_{21} K A_{12} , \tag{7}$$

where K is the diagonal matrix defined by

$$K_{ii} = \begin{cases} (A_{11} \mathbf{e}_1)_i^{-1} & \text{if } (A_{11} \mathbf{e}_1)_i \neq 0 \\ 0 & \text{otherwise} , \end{cases} \tag{8}$$

$\mathbf{e}_1 = (1\,1\ldots1)^T$ being the vector with all components equal to unity.

From a practical point of view, note that this method is pretty easy to implement, and that if A originates from a regular discretization, S will generally present relatively to the coarse grid a structure similar to that of A on the fine grid. From a more theoretical point of view, a key property is that when A is an M–matrix with nonnegative row-sum, then S itself is an M–matrix with nonnegative row-sum [22]. Hence a recursive use of the preconditioning technique is coherent in this context.

Now, these nice properties would be useless without a good clustering of the eigenvalues of $S^{-1}S_A$. In [21], considering five point approximations of 2D self-adjoint second order PDEs, we show that the eigenvalues, necessarily real, are bounded below by 1 and above by 2 when the PDE coefficients are piecewise constant on the coarse mesh. This result is partly based on the observation that, under the given assumptions, S is equal to the coarse grid discretization matrix times some scaling factor.

Now, in general, S may have little in common with a coarse discretization and the analysis of the spectrum of $S^{-1}S_A$ has to be based on different arguments. When the eigenvalues are complex, the most viable way is based on so-called Fourier analysis (see e.g. [18]). The latter comes with some limitations, namely one is restricted to matrices corresponding to a constant stencil acting on a regular periodic grid. In the discrete PDE context, this means that one has to assume that the PDE coefficients are constant and that the boundary conditions are periodic (More precisely, since this is in most cases non physical, one presumes that the quality of the preconditioner is essentially independent of the type of boundary conditions, which generally turns out to be true for the family of methods considered here).

Within this setting an analysis is possible because S will correspond to a constant stencil acting on the (periodic) coarse grid, and hence will share a common set of eigenvectors with S_A. For instance, for a 2D grid with $2N \times 2N$ nodes defined on $\Omega =]-1,1] \times]-1,1]$, these eigenvectors are the $4 N^2$ functions

$$v_{k\ell} = e^{i \pi k x} e^{i \pi \ell y}, \qquad 1 - N \leq k, \ell \leq N$$

evaluated at the (coarse) grid points. Moreover, analytic expressions for the corresponding eigenvalues of both S and S_A are easily deduced, see [18, Section 8.1.2] and [24] for details. Taking their ratio, one then straightforwardly gets an explicit expression for the eigenvalues of $S^{-1}S_A$, and one may readily obtain an idea of the quality of the eigenvalue distribution by plotting them as done on Fig. 1 for two examples of 2D stencils.

It is more tedious to prove a domain of confidence for the eigenvalues because of the many parameters involved. Nevertheless, we succeeded to prove that, whatever the 2D five point stencil with zero row-sum and nonpositive offdiagonal entries, any eigenvalue λ of $S^{-1}S_A$ satisfies

$$\left| \lambda^{-1} - 1 - i c \right| \leq \frac{1}{2} \tag{9}$$

for some real constant c such that $|c| \leq \frac{1}{4}$ [22]. Note that this result holds independently of the grid size.

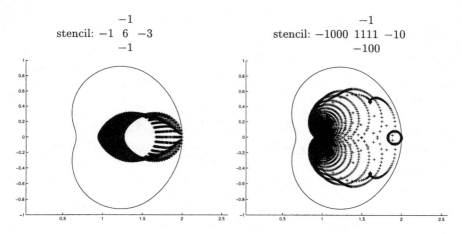

Fig. 1. Spectrum of the preconditioned system for some examples of constant five point stencil (100×100 periodic grid); the continuous line delimits the region that contains the eigenvalues according to (9).

3   Practical Two-Level Schemes

If, from a theoretical point of view, it is comfortable to assume the exact inversion of A_{11}, in practice, two such inversions are required for each application of the preconditioner (5), so that the overall cost will be kept reasonable only if the number of operations involved remains a small multiple of the number of nonzero elements in A_{11}. This means that practical two-level schemes have to rely on an approximate inversion of A_{11}. The preconditioner has thus to be rewritten

$$ B = \begin{pmatrix} P \\ A_{21} & S \end{pmatrix} \begin{pmatrix} I & P^{-1}A_{12} \\ & I \end{pmatrix} , \tag{10}$$

where P is a preconditioner for A_{11}, possibly implicitly defined by a very few steps of some iterative procedure.

Here, we have to recall that with a naive choice of P, eigenvalues of $B^{-1}A$ may be unbounded even though those of $P^{-1}A_{11}$ and of $S^{-1}S_A$ are both nicely clustered [19]. This is perfectly illustrated with the numerical results from [19] that are reproduced in Table 1. They concern the standard five point approximation of the Poisson equation in the unit square, using a uniform grid of mesh size h. For both standard ILU and *Modified* ILU (MILU) (e.g. [3,16]) preconditioning of A_{11}, we give the extremal eigenvalues and the condition number for $P^{-1}A_{11}$ on the one hand and $B^{-1}A$ on the other hand, whenever taking S equal to the standard coarse grid discretization matrix. One sees that both approximate factorizations are here of similar quality as

h^{-1}	$P^{-1}A_{11}$			$B^{-1}A$		
	λ_{min}	$\lambda_{\max}$	κ	λ_{min}	$\lambda_{\max}$	κ
MILU preconditioning of A_{11}						
16	1.00	1.20	1.20	0.51	1.25	2.45
32	1.00	1.21	1.21	0.50	1.27	2.54
64	1.00	1.21	1.21	0.50	1.29	2.58
128	1.00	1.21	1.21	0.50	1.29	2.58
ILU preconditioning of A_{11}						
16	0.88	1.09	1.24	0.510	1.42	2.78
32	0.88	1.09	1.24	0.380	2.28	6.00
64	0.87	1.09	1.25	0.176	4.97	28.30
128	0.87	1.09	1.25	0.058	15.00	258.00

Table 1. Results for the model Poisson problem

preconditioner of A_{11} alone. However, remembering that $\lambda_{\min}(B^{-1}A) = \frac{1}{2}$ and $\lambda_{\max}(B^{-1}A) = 1$ when $P = A_{11}$ [21], MILU preconditioning of A_{11} implies only a very moderate degradation, whereas using the standard ILU preconditioning of A_{11} has disastrous effects on the eigenvalue distribution of $B^{-1}A$.

In [19], we develop a proper analysis of this phenomenom for the case of symmetric and positive definite matrices. We show that the condition number of $B^{-1}A$ is stable when P, besides being symmetric and positive definite, satisfies the two following algebraic requirements: $\forall \mathbf{v} \in \mathcal{C}^n$,

$$\mathbf{v}^T P \mathbf{v} \leq \mathbf{v}^T A_{11} \mathbf{v} \tag{11}$$

and

$$\mathbf{v}^T \left(A_{22} - A_{21}P^{-1}A_{12}\right) \mathbf{v} \geq \xi \, \mathbf{v}^T S_A \mathbf{v} \tag{12}$$

for some $\xi < 1$ (possibly negative).

For instance, when (12) holds with $\xi = 0$, eigenvalue bounds are obtained that yield at least (simplifying somewhat the expressions for the sake of clarity)

$$\frac{1}{2}\lambda_{\min}\left(S^{-1}S_A\right) \leq \lambda \leq 2\lambda_{\max}\left(P^{-1}A_{11}\right)\lambda_{\max}\left(S^{-1}S_A\right) \tag{13}$$

for all $\lambda \in \sigma\left(B^{-1}A\right)$.

Now, (12) holds with $\xi = 0$ if and only if

$$\begin{pmatrix} P & A_{12} \\ A_{21} & A_{22} \end{pmatrix} \tag{14}$$

is nonnegative definite. Thus, to satisfy both this and (11), one has to find a proper preconditioner of A_{11} that is less than A_{11} in the positive definite sense, but remains nevertheless sufficiently positive definite to maintain the global positive definiteness of (14).

One may therefore ask whether this is at all possible. We have no general answer, but the case of symmetric M–matrices with nonnegative row-sum is analyzed in detail in [19,21]. There, we first observe that A_{11} is then itself always a symmetric M–matrix with nonnegative row-sum, and hence that *Modified* ILU factorizations of A_{11} are well defined. Further, it is well known that (11) holds for such factorizations, whereas we show that, whatever the level of fill-in, P is not less than Δ in the positive definite sense, where Δ is as above the diagonal matrix with same row-sum as A_{11}. Using standard properties of M–matrices, it is then relatively easy to show that (14) is non-negative definite, hence that (12) holds with $\xi = 0$, see [19] for details.

In [21], we even further refine this analysis and show that, whenever using the approximate Schur complement S described in Sect. 2, one has

$$\lambda_{\min}\left(B^{-1}A\right) \geq 1 \,,$$
$$\lambda_{\max}\left(B^{-1}A\right) \leq \lambda_{\max}\left(P^{-1}A_{11}\right) \cdot \lambda_{\max}\left(S^{-1}S_A\right) \,.$$

On the other hand, considering five point finite difference matrices arising from self-adjoint 2D second order elliptic PDEs, we moreover show that $\lambda_{\max}\left(P^{-1}A_{11}\right)$ cannot exceed $\frac{3}{2}$ even with the simplest scheme of a MILU factorization without fill-in (provided the PDE coefficients are piecewise constant on the coarse mesh). Hence these theoretical results show that the situation seen on Table 1 is not proper to that example, and further that the behavior with MILU preconditioning is likely to be even better when using the algebraic approximate Schur complement introduced in Sect. 2.

Note that, without any fill-in, this MILU factorization of A_{11} is in fact very cheap, and, clearly, such an approach can be much better, in term of cost efficiency, than the ones that attempt to mimic the exact inversion of A_{11} by performing several inner iterations.

Concerning non symmetric M–matrices with nonnegative row-sum, MILU factorization are still well defined [3, Section 7.1]. For completeness, let us recall that, discarding all fill-in, P writes

$$P = (Q - E)Q^{-1}(Q - F)$$

where $(-E)$ is the strictly lower part of A_{11}, $(-F)$ is the strictly upper part of A_{11} and Q is diagonal and computed such that

$$P\,\mathbf{e}_1 \;=\; A_{11}\,\mathbf{e}_1 \,.$$

As in the symmetric case, the good conditioning properties of A_{11} together with the robustness of ILU methods reflect in the eigenvalues of $P^{-1}A_{11}$, that are therefore very well clustered. So, the main danger comes again from the global effect on the eigenvalues of $B^{-1}A$. Since an extension of the theoretical results for the symmetric case is till now out of reach, we assessed in [22] the corresponding eigenvalue distribution through extensive numerical experiments. More precisely, we considered a wide variety of constant five point stencils acting on a regular 128×128 grid with Dirichlet boundary conditions, and we computed the asymptotic convergence factor ρ associated to fixed point iterations of the form

$$\mathbf{u}_{k+1} = \mathbf{u}_k + \tau^{-1} B^{-1} (\mathbf{b} - A \mathbf{u}_k) \tag{15}$$

with $\tau = 1.5$ (minimum 200 iterations). Since

$$\rho = \max_{\lambda \in \sigma(B^{-1}A)} \left| \frac{\lambda}{\tau} - 1 \right|,$$

was at worst equal to 0.55, this means that all eigenvalues are inside a circle of center $(1.5, 0)$ and radius 0.825. We therefore concluded that the eigenvalue distribution is in this case not too deeply perturbed from that of the "ideal" two-level scheme.

The result of this computation is also illustrated on Fig. 2 for the two example stencils already considered in Fig. 1. Note that the circle does not necessarily contains all eigenvalues for the "ideal" preconditioner because the latter were computed assuming periodic boundary conditions, and also because MILU preconditioning of A_{11} tends anyway to shift all eigenvalues somewhat to the right.

4   Multilevel Schemes

The two-level preconditioner (10) requires solving a system $S \mathbf{x}_2 = \mathbf{y}_2$. In general, this system remains too large to make direct solvers competitive, and iterative procedures have to be considered. Here again, for cost efficiency reasons, only approximate solves are performed, and the preconditioner has to be rewritten

$$B = \begin{pmatrix} P & \\ A_{21} & M^{-1} \end{pmatrix} \begin{pmatrix} I & P^{-1}A_{12} \\ & I \end{pmatrix}, \tag{16}$$

where M stands for the used approximate inverse of S. Note that we can here not meet stability problems like those mentioned in Sect. 3. For instance, in the symmetric case, it can be proved that the condition number is at worst multiplied by $\kappa(M S) = \lambda_{\max}(M S)/\lambda_{\min}(M S)$ (assuming M is symmetric and positive definite).

Now, several choices are possible to define this approximate inverse.

Firstly, one may use inner iterations with a standard preconditioner, if the grid size is sufficiently reduced to make the latter competitive [4]. Note that if a Krylov subspace method is used to accelerate the convergence of

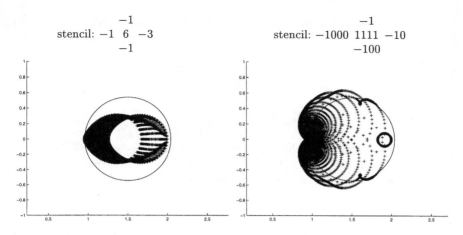

stencil:
$$\begin{matrix} & -1 & \\ -1 & 6 & -3 \\ & -1 & \end{matrix}$$

stencil:
$$\begin{matrix} & -1 & \\ -1000 & 1111 & -10 \\ & -100 & \end{matrix}$$

Fig. 2. Spectrum of $B^{-1}A$ with exact inversion of A_{11} (using periodic boundary conditions), and circle containing the eigenvalues whenever using a MILU factorization to approximate A_{11} for the problem with Dirichlet boundary conditions (128×128 grid).

these inner iterations, then the global preconditioner (16) is in fact slightly variable from one iteration to the next, and this should be taken into account in the choice of the outer Krylov accelerator (see [27] and [23] for "flexible" variants of respectively GMRES and the conjugate gradient method).

However, most works prefer to focus on *multilevel* schemes, in which a preconditioner for S is defined by applying the same two-level procedure (16) that was used to precondition the system matrix A. This approach is then followed recursively until the number of nodes left is fairly small, allowing a cheap exact factorization of the remaining subsystem.

Now, within this framework, there are still several possible strategies.

1. One may approximate the inverse of S just by the inverse of its so defined preconditioner. This approach, referred to as V cycle (by analogy with multigrid [18]), is quite simple, but the condition number is then likely to grow exponentially with the number of levels, except whenever working with so-called hierarchical finite element spaces [33].

2. One therefore mostly uses

$$M = \mathcal{P}_m \left(B^{(S)^{-1}} S \right) S^{-1} , \tag{17}$$

where $B^{(S)}$ is the preconditioner for S and $\mathcal{P}_m$ is a polynomial such that $\mathcal{P}_m(0) = 0$. This means in practice that one uses a few steps of an iterative method to solve approximately the system with S:

$$
\begin{aligned}
\mathbf{x}_2^{(0)} &= 0, \\
\mathbf{x}_2^{(k+1)} &= \mathbf{x}_2^{(k)} + \tau_k^{-1} \, B^{(S)\,-1} \left(\mathbf{y}_2 - S\,\mathbf{x}_2^{(k)} \right), \qquad k = 0, \ldots, m-1 .
\end{aligned}
\tag{18}
$$

One speaks here of W cycle.

In the symmetric case, using for instance shifted Chebyshev polynomials, one may then prove that the condition number is stabilized independently of the number of levels under quite reasonable assumptions. More precisely, the stabilization is obtained provided that

$$
\overline{\kappa}_2 \leq m^2 , \tag{19}
$$

where $\overline{\kappa}_2$ is an upper bound for the condition number of the two-level method. The drawback of this approach is that the parameters are to be determined in advance according some eigenvalue estimates, and if the estimation is too optimistic, one easily ends up with an indefinite preconditioner. A robust approach consists then in computing these eigenvalue estimates at run time by a few steps of an iterative eigensolver [9], but this seems feasible only in the symmetric case.

3. In [20] we develop another approach, in which one adds to a simple V cycle as referred above some smoothing iterations in a much similar way as it is done in standard multigrid methods (Except that smoothing is not required on the finest grid since we seek only to define an approximate inverse of the successive coarse grid matrices). With a proper scaling of the approximate Schur complements, this method offers a stabilization of the condition number without requiring the a priori knowledge of some parameters. This provides therefore some additional robustness, but which comes with the limitation that this approach can be effective only in contexts for which (at least some) more standard multigrid algorithms would also work. On the other hand, as highlighted in [22], W cycle based algebraic multilevel methods have the nice property that they do not require smoothing, and can therefore be effective in contexts where it is intrinsically difficult to define a "robust" smoother.

4. Finally, it is possible to design a parameter free W cycle variant by exchanging the iteration (18) for inner iterations with a proper Krylov subspace accelerator [27,23]. Within this context, it is much more difficult to *prove* that the condition number is stabilized at a reasonable cost, due to the variations in the preconditioner. However, according the results in [12,23], this approach, which deserves to be investigated further, is expected to be quite robust in practice.

Computational Complexity For each application of the multilevel preconditioner, the work on a given level k is proportional to the number of unknowns n_k in that level multiplied by the number of times this level is visited. The latter number is just 1 for V cycle variants, whereas it is equal to m^k if m inner iterations are performed, or, equivalently, if the polynomial that defines M has degree m (here, we have assumed that the finest grid corresponds to $k = 0$, the next coarser one to $k = 1$, etc). Thus, for the global work, assuming $n_k \approx \sigma^{-k} n$ (each level reduction divides the number of unknowns by approximately σ), one has

$$W \sim \sum_{k=0}^{\ell} \left(\frac{m}{\sigma}\right)^k = \frac{1 - \left(\frac{m}{\sigma}\right)^{\ell+1}}{1 - \frac{m}{\sigma}}$$

and a reasonable bound is obtained as long as $m < \sigma$.

This means that a stabilization of the condition number cannot be obtained unconditionally, since m has to be adjusted on the other hand in function of the conditioning properties for the two-level case (see (19) in the symmetric case).

This is however not a strong limitation. For instance, in 2D examples with geometric coarsening, we have seen that $\bar{\kappa}_2$ was slightly larger than 2 whereas $\sigma = 4$, thus both requirements are perfectly matched with $m = 2$. In [22], quite satisfactory results are also obtained with this value in the non symmetric case. It may even be observed that the resulting preconditioner is then in fact quite cheap. Detailed evaluation shows indeed that each solve with B requires, for five point matrices, not more than $36\,n$ basic floating point operations (*flops*). The global cost for one preconditioner solve plus one multiplication by A is thus only about $45\,n$ *flops*, that is roughly only twice the cost that one would have with a mere ILU(0) like preconditioner.

5   Numerical results

To illustrate the efficiency of algebraic multilevel preconditioning, we reproduce on Fig. 3 some numerical results from [21], where a basic multilevel method is compared to standard (block) incomplete LU decompositions. The problem under consideration is the model Poisson problem on the unit square discretized by five point central finite difference, using a uniform mesh of mesh size h; all considered preconditioners are combined with the conjugate gradient acceleration, using the zero vector as initial approximation.

The multilevel method uses on each level the Schur complement approximation described in Sect. 2 and a Modified ILU factorization (without fill-in) to approximate A_{11}, as discussed in Sect. 3; on all intermediate levels, the approximation M to S^{-1} is defined by the iteration (18) with $m = 2$ and $\tau_0 = \tau_1 = \sqrt{\frac{8}{3}}$.

One sees that the multilevel method has grid independent convergence and already outmatches standard preconditioning techniques for fairly moderate

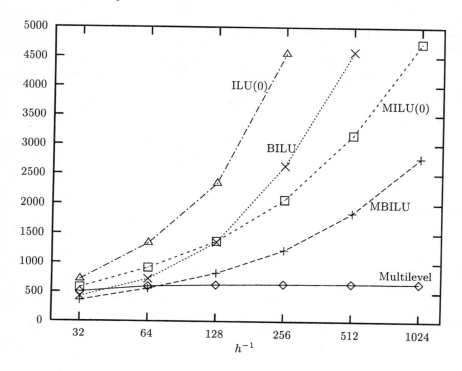

Fig. 3. 2D Poisson problem: for different preconditioning methods, number of basic floating point operations per unknown needed to reduce the relative residual error by 10^{-6}.

problem sizes. This has to be related to the fact that the considered multilevel scheme is rather cheap for such an high quality preconditioning.

Many other numerical results are reported in [20–22]. They essentially display the high robustness of the proposed preconditioning approach: not only is the convergence independent of mesh refinement, but also it only weakly varies with problem parameters such as the discontinuity or anisotropy ratios in the PDE coefficients, the level of stretching in case on non uniform meshes, and the Reynolds number in convection–diffusion problems.

References

1. Axelsson, O.: On multigrid methods of the two-level type. Multigrid Methods (W. Hackbusch, U Trottenberg, eds.), Lectures Notes in Mathematics No. 960, Springer-Verlag (Berlin Heidelberg New York) 1981, pp. 352–367.
2. Axelsson, O.: The method of diagonal compensation of reduced matrix entries and multilevel iteration. J. Comput. Appl. Math. **38** (1991) 31–43.
3. Axelsson, O.: Iterative Solution Methods. University Press, Cambridge, 1994.

4. Axelsson, O.: The stabilized v–cycle method. J. Comput. Appl. Math. **74** (1996) 33–50.
5. Axelsson, O., Barker, V. A.: Finite Element Solution of Boundary Value Problems. Theory and Computation. Academic Press, New York, 1984.
6. Axelsson, O., Eijkhout, V.: Analysis of recursive 5-point/9-point factorization method. Preconditioned Conjugate Gradient Methods (Axelsson, O., Kolotilina, L.Yu., eds.), Lectures Notes in Mathematics No. 1457, Springer-Verlag (Berlin Heidelberg New York), 1990, pp. 154–173.
7. Axelsson, O., Eijkhout, V.: The nested recursive two level factorization for nine-point difference matrices. SIAM J. Sci. Stat. Comput. **12** (1991) 1373–1400.
8. Axelsson, O., Gustafsson, I.: Preconditioning and two-level multigrid methods of arbitrary degree of approximation. Math. Comp. **40** (1983) 214–242.
9. Axelsson, O., Neytcheva, M.: Algebraic multilevel iterations for Stieltjes matrices. Numer. Lin. Alg. Appl. **1** (1994) 213–236.
10. Axelsson, O., Vassilevski, P. S.: Algebraic multilevel preconditioning methods, I. Numer. Math. **56** (1989) 157–177.
11. Axelsson, O., Vassilevski, P. S.: Algebraic multilevel preconditioning methods, II. SIAM J. Numer. Anal. **27** (1990) 1569–1590.
12. Axelsson, O., Vassilevski, P. S.: Variable-step multilevel preconditioning methods. I. selfadjoint and positive definite elliptic problems. Numer. Lin. Alg. Appl. **1** (1994) 75–101.
13. Bank, R. E.: Hierarchical bases and the finite element method. Acta Numerica **5** (1996) 1–43.
14. Bank, R. E., Dupont, T. F.: Analysis of a two-level scheme for solving finite element equations. Tech. Report CNA-159, Center for Numerical Analysis, The University of Texas at Austin, Texas, USA, 1980.
15. Bank, R. E., Dupont, T. F., Yserentant, H.: The hierarchical basis multigrid method. Numer. Math. **52** (1988) 427–458.
16. Chan, T. F., van der Vorst, H. A.: Approximate and incomplete factorizations. Parallel Numerical Algorithms (David E. Keyes, Ahmed Samed, V. Venkatakrishnan, eds.), ICASE/LaRC Interdisciplinary Series in Science and Engineering, Volume 4, Kluwer Academic (Dordecht), 1997, pp. 167–202.
17. Greenbaum, A.: Iterative Methods for Solving Linear Systems. Frontiers in Applied Mathematics, vol. 17, SIAM, Philadelphia, PA, 1977.
18. Hackbusch W.: Multi-Grid Methods and Applications. Springer, Berlin, 1985.
19. Notay, Y.: Using approximate inverses in algebraic multilevel methods. Numer. Math. **80** (1998) 397–417.
20. Notay, Y.: Optimal V cycle algebraic multilevel preconditioning. Numer. Lin. Alg. Appl. **5** (1998) 441–459.
21. Notay, Y.: Optimal order preconditioning of finite difference matrices. To appear in SIAM J. Sci. Comput., 1999.
22. Notay, Y.: A robust algebraic preconditioner for finite difference approximations of convection-diffusion equations. Tech. Report GANMN 99-01, Université Libre de Bruxelles, Brussels, Belgium, 1999.
23. Notay, Y.: Flexible conjugate gradient. Tech. Report GANMN 99-02, Université Libre de Bruxelles, Brussels, Belgium, 1999.
24. Reusken, A.: Fourier analysis of a robust multigrid method for convection-diffusion equations. Numer. Math. **71** (1995) 365–398.
25. Reusken, A.: A multigrid method based on incomplete Gaussian elimination. Numer. Lin. Alg. Appl. **3** (1996) 369–390.

26. Reusken, A.: On a robust multigrid solver. Computing **56** (1996) 303–322.
27. Saad, Y.: Iterative Methods for Sparse Linear Systems. PWS publishing, New York, 1996.
28. Saint-Georges, P., Warzee, G., Beauwens, R, Notay, Y.: High performance PCG solvers for FEM structural analyses. Int. J. Numer. Meth. Engng. **39** (1996) 1313–1340.
29. Stüben, K.: Algebraic multigrid (AMG): an introduction with applications. Multigrid (U. Trottenberg, C. Oosterlee, A. Scüller, eds.), Academic Press, 1999, To appear.
30. van der Ploeg, A, Botta, E, Wubs, F.: Nested grids ILU-decomposition (NGILU). J. Comput. Appl. Math. **66** (1996) 515–526.
31. Vassilevski, P. S.: Nearly optimal iterative methods for solving finite element elliptic equations based on the multilevel splitting of the matrix. Tech. Report # 1989-09, Institute for Scientific Computation, University of Wyoming, Laramie, USA, 1989.
32. Vassilevski, P. S.: Hybrid V-cycle algebraic multilevel preconditioners. Math. Comp. **58** (1992) 489–512.
33. Vassilevski, P. S.: On two ways of stabilizing the hierarchical basis multilevel methods. SIAM Review **39** (1997) 18–53.
34. Wagner C., Kinzelbach, W., Wittum, G.: Schur-complement multigrid. Numer. Math. **75** (1997) 523–545.
35. Yserentant, H.: On the multi-level splitting of finite element spaces. Numer. Math. **49** (1986) 379–412.

On Algebraic Multilevel Preconditioners in Lattice Gauge Theory

Björn Medeke

Department of Mathematics
Institute of Applied Computer Science
University of Wuppertal
Gaußstraße 20, D-42097 Wuppertal, Germany

Abstract. Based on a Schur complement approach we develop a parallelizable multi-level preconditioning method for computing quark propagators in lattice gauge theory.

1 Introduction

Lattice gauge theory (LGT) is a discretization of quantum chromodynamics which is generally accepted to be the fundamental physical theory of strong interactions among the quarks as constituents of matter.

The most time-consuming part of a numerical simulation in lattice gauge theory with Wilson fermions on the lattice is the computation of quark propagators within a chromodynamic background gauge field. Mathematically, quark propagators are obtained by solving the inhomogenous lattice-Dirac equation

$$A\psi = \varphi \tag{1}$$

These computations use up a major part of the world's high performance computing power.

After reviewing the Wilson fermion matrix in Sect. 2, we present efficient Krylov subspace methods for the computation of quark propagators in Sect. 3, introduce the basic idea of multilevel preconditioning and discuss multilevel preconditioners for the Wilson fermion matrix. In Section 4, for a simple model problem numerical results are presented. Finally, in Sect. 5, we draw some conclusions.

2 Wilson Fermion Matrix

To fix notation, we briefly introduce the Wilson fermion matrix and recall some basic properties. For a more detailed introduction, see [13], [12] and the references therein.

2.1 Structure of the Wilson Fermion Matrix

For the Wilson fermion action, the Wilson fermion matrix $A = I - \kappa D$ describes a nearest neighbour coupling with periodic boundary conditions on a four-dimensional regular hypercubic space-time lattice with lattice sites

$$\Omega = \left\{ x = (x_1, x_2, x_3, x_4) \mid x_i = 1, \ldots, n_i, \quad n_i = 2^{N_i} \right\}. \qquad (2)$$

Th off-diagonal part κD of A depends on a real non-negative parameter κ, the standard Wilson hopping parameter, and the so-called hopping matrix D,

$$D_{x,y} = \sum_{\mu=1}^{4} \left((I_4 - \gamma_\mu) \otimes U_\mu(x) \right) \delta_{x,y-e_\mu}$$

$$+ \sum_{\mu=1}^{4} \left((I_4 + \gamma_\mu) \otimes U_\mu^H(x - e_\mu) \right) \delta_{x,y+e_\mu}. \qquad (3)$$

Here, the complex 4×4 matrices $I_4 \pm \gamma_\mu$ are projectors onto two-dimensional subspaces and the 3×3 matrices $U_\mu(x)$ are from $SU(3)$. The configuration $\{U_\mu(x)\}$ represents the background gauge field, $\{U_\mu(x)\}$ is labeled cold (hot), if $U_\mu(x) = I_3$ ($U_\mu(x)$ randomly choosen). For realistic configurations, the matrices $U_\mu(x)$ are generated stochastically.

Due to the Kronecker product structure of the coupling terms it is clear that there exist 12 unknowns on each lattice site. Thus, with $n = n_1 n_2 n_3 n_4$ denoting the number of lattice sites, $12n$ is the total system size.

2.2 Basics Properties of the Wilson Fermion Matrix

Basically, there are two important symmetries worthwhile to mention. First, the hopping matrix D is a γ_5-symmetric[1] matrix,

$$\Gamma_5 D = D^H \Gamma_5, \quad \Gamma_5 = I_n \otimes (\gamma_5 \otimes I_3) \qquad (4)$$

where $\gamma_5 = [e_3|e_4|e_1|e_2]$ is just a simple permutation matrix. Second, it is odd even symmetric, i.e. if we order all odd sites before the even sites (odd-even or red-black ordering), the hopping matrix D becomes

$$D = \begin{pmatrix} 0 & D_{oe} \\ D_{eo} & 0 \end{pmatrix}. \qquad (5)$$

Due to the γ_5-symmetry (4) of D the eigenvalues of D come in complex conjugate pairs λ and $\bar{\lambda}$. The odd-even symmetry (5) implies that the eigenvalues of D appear in pairs λ and $-\lambda$.

Figure 1 shows a typical eigenvalue distribution of the Wilson fermion matrix A. Obviously, for all $0 \leq \kappa < \kappa_c$ the Wilson fermion matrix A is positive real, i.e. all eigenvalues lie in the right half plane.

---

[1] more correctly: γ_5-hermitian

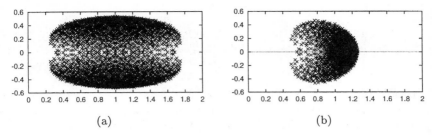

Fig. 1. Eigenvalue distribution of Wilson fermion matrix A (a) and corresponding odd-even reduced system A_e (b) for a realistic configuration, i.e. $U_\mu(x)$ are generated stochastically, with a moderate hopping parameter $\kappa \ll \kappa_c$.

For $\kappa \approx \kappa_c$ the solution is physically significant. This choice corresponds to the computation of quark propagators in the critical regime of small quark masses on the lattice,

$$m_q = \frac{1}{2a}\left(\frac{1}{\kappa} - \frac{1}{\kappa_c}\right),\tag{6}$$

where a is the lattice spacing.

3   Efficient Krylov Subspace Solvers

With the development of Krylov subspace methods, now efficient iterative solvers for linear systems are widely available. An advisable survey of iterative methods is given in [3], for implications for LGT solvers we refer to [10], [5] and the references therein.

In the case of Wilson fermions, experiments with full GMRES [18] indicate that BiCGStab [19], BiCG [8] and QMR[2] [9] are already almost optimal for Krylov subspaces generated by A, see [10].

Further improvements in Wilson fermion solvers are only to be expected via preconditioning. Hence, the development of efficient preconditioners is crucial. For instance, SSOR preconditioning of fermion matrix inversions which is parallelized using a locally lexicographic lattice sub-division has been shown to be very efficient for standard Wilson fermions [7] as well as for more complicated fermion actions [6], [4].

In this section, after reviewing the standard odd-even preconditioner, we focus our attention on algebraic multilevel preconditioning [1], [2], [16], [17].

3.1   Odd Even Preconditioning

When writing down (1) one has the freedom of choosing any ordering scheme for the lattice sites. Using the standard odd-even ordering and rewriting (1)

---
[2] In Lanczos-based methods like QMR and BiCG the γ_5-symmetry can be exploited, see [5].

yields

$$\begin{pmatrix} I & -\kappa D_{oe} \\ -\kappa D_{eo} & I \end{pmatrix} \begin{pmatrix} \psi_o \\ \psi_e \end{pmatrix} = \begin{pmatrix} \varphi_o \\ \varphi_e \end{pmatrix} \tag{7}$$

which can be reduced to half its size by solving

$$A_e \psi_e = \varphi_e + \kappa D_{eo}\varphi_o, \quad A_e = I - \kappa^2 D_{eo}D_{oe} \tag{8}$$

on the even sites, only. Once this odd even reduced system is solved, the odd sites of ψ can be obtained by a simple substitution process,

$$\psi_o = \varphi_o + \kappa D_{oe}\psi_e. \tag{9}$$

Obviously, odd even preconditioning can be easily implemented. Due to the favorable eigenvalue distribution, cf. Fig. 1, in comparison to the original system an improvement of a factor between 2 and 3 both in iteration numbers and CPU time can be observed. Therefore, it has become standard to solve the odd-even reduced system.

3.2   Algebraic Multilevel Preconditioning

In a more general framework, we now consider the decomposition $\Omega = \Omega_f \cup \Omega_c$ of all lattice sites Ω into two disjoint sets Ω_c and $\Omega_f = \Omega \backslash \Omega_c$. Here, the subscripts c and f denote coarse and fine sites, respectively.

Assume that the Wilson fermion matrix is partitioned in a 2×2 block form

$$A = \begin{pmatrix} A_{ff} & A_{fc} \\ A_{cf} & A_{cc} \end{pmatrix} \tag{10}$$

where the first block of unknowns corresponds to the fine sites Ω_f and the second block of unknowns to the coarse sites Ω_c.

To construct a two-level preconditioner we remark that an exact factorization of A yields

$$A^{-1} = \begin{pmatrix} I & -A_{ff}^{-1}A_{fc} \\ 0 & I \end{pmatrix} \begin{pmatrix} A_{ff}^{-1} & 0 \\ 0 & S_A^{-1} \end{pmatrix} \begin{pmatrix} I & 0 \\ -A_{cf}A_{ff}^{-1} & I \end{pmatrix} \tag{11}$$

with $S_A = A_{cc} - A_{cf}A_{ff}^{-1}A_{fc}$ being the Schur complement A/A_{ff}.

Suppose that the matrices A_{ff}^{-1} and S_A are known explicitly and ψ and φ are partitioned accordingly, the original linear system

$$\begin{pmatrix} A_{ff} & A_{fc} \\ A_{cf} & A_{cc} \end{pmatrix} \begin{pmatrix} \psi_f \\ \psi_c \end{pmatrix} = \begin{pmatrix} \varphi_f \\ \varphi_c \end{pmatrix} \tag{12}$$

can be solved immediately via

$$S_A\psi_c = \varphi_c - A_{cf}A_{ff}^{-1}\varphi_f, \quad A_{ff}\psi_f = \varphi_f - A_{fc}\psi_c. \tag{13}$$

The described approach is essentially block Gaussian elimination which is not feasible in general, since A_{ff}^{-1} and the Schur complement S_A are dense.

Therefore, both matrices have to be replaced by suitable approximations. Then, a two-level preconditioner M of A is defined by

$$M^{-1} = \begin{pmatrix} I & -\widetilde{A}_{ff}^{-1}A_{fc} \\ 0 & I \end{pmatrix} \begin{pmatrix} \widetilde{A}_{ff}^{-1} & 0 \\ 0 & \widetilde{S}_A^{-1} \end{pmatrix} \begin{pmatrix} I & 0 \\ -A_{cf}\widetilde{A}_{ff}^{-1} & I \end{pmatrix}. \tag{14}$$

In (14), the matrices R and P,

$$R = \begin{pmatrix} -A_{cf}\widetilde{A}_{ff}^{-1} & I \end{pmatrix}, \quad P = \begin{pmatrix} -\widetilde{A}_{ff}^{-1}A_{fc} \\ I \end{pmatrix}, \tag{15}$$

may be interpreted as matrix dependent restriction and matrix dependent prolongation, respectively, whereas the Schur complement approximation $\widetilde{S}_A$ can be considered as a coarse grid matrix.

To get the multi-level case, we apply recursively the two-level procedure to the Schur complement approximation $\widetilde{S}_A$.

We conclude the description of the basic multi-level preconditioning idea with a short note about odd-even preconditioning which can be regarded as a two-level preconditioner, where Ω_c consists of odd sites and Ω_f of even sites, only. Since odd sites couple only with even sites and vice versa, the matrices A_{ff} and A_{cc} both are diagonal. Thus, the restriction and the prolongation are optimal, i.e. the approximate inverse is in fact the exact one. The resulting two-level preconditioner is ideal in the sense that once the exact coarse grid solution is available, the original problem is solved, see (13). Clearly, the exact Schur complement $S_A = I - \kappa^2 D_{eo}D_{oe}$ suffers from fill-in, because the Schur complement does not only represent a nearest but additionally a next-to-nearest neighbour coupling on the coarse sites.

In the following, for simplicity, we restrict ourselves to the two-level case.

Fine/Coarse Partitioning For the selection of the coarse sites we make use of the geometric information about the regular structure of the underlying hypercubic space-time lattice Ω.

Let L_i be the set of lattice sites x with exactly i even components. Then, Ω is decomposed into five mutually disjoint sets L_i,

$$\Omega = \bigcup_{i=0}^{5} L_i, \quad L_i \cap L_j = \emptyset, \quad i \neq j. \tag{16}$$

Obviously, each lattice site $x \in L_1 \cup L_3$ is odd, while each site $x \in L_0 \cup L_2 \cup L_4$ is even. As coarse sites we select

$$\Omega_c = L_0 \cup L_4, \quad |\Omega_c| = \frac{1}{8}n. \tag{17}$$

For this choice, each coarse site $x \in L_0$ (L_4) couples with 8 fine sites $y \in L_1$ (L_3), each fine site $x \in L_1$ (L_3) with both 6 fine sites $y \in L_2$ and 2 ones $y \in L_0$ (L_4) and, finally, each fine site $x \in L_2$ couples with 4 fine sites $y \in L_1$ and also with 4 sites $y \in L_3$.

Hence, using a fine-coarse ordering, see Fig. 2, we obtain

$$A = \begin{pmatrix} A_{ff} & A_{fc} \\ A_{cf} & A_{cc} \end{pmatrix} = I - \kappa \begin{pmatrix} D_{ff} & D_{fc} \\ D_{cf} & 0 \end{pmatrix} \tag{18}$$

and, accordingly, $\Gamma_5 = \mathrm{diag}(\Gamma_5^{(f)}, \Gamma_5^{(c)})$ with

$$\Gamma_5^{(f)} = I_{\frac{7}{8}n} \otimes (\gamma_5 \otimes I_3), \quad \Gamma_5^{(c)} = I_{\frac{1}{8}n} \otimes (\gamma_5 \otimes I_3). \tag{19}$$

Furthermore, we suppose that the fine and coarse grid sites are given in the order $L_1, L_3,\ L_2,\ L_0$ and L_4. Then, the sub-matrices of D are partitioned as follows:

$$D_{ff} = \begin{pmatrix} 0 & 0 & D_{L_1,L_2} \\ 0 & 0 & D_{L_3,L_2} \\ D_{L_2,L_1} & D_{L_2,L_3} & 0 \end{pmatrix}, \tag{20}$$

$$D_{fc} = \begin{pmatrix} D_{L_1,L_0} & 0 \\ 0 & D_{L_3,L_4} \\ 0 & 0 \end{pmatrix}, \quad D_{cf} = \begin{pmatrix} D_{L_0,L_1} & 0 & 0 \\ 0 & D_{L_4,L_3} & 0 \end{pmatrix}. \tag{21}$$

Here, D_{L_i,L_j} represents the couplings of lattice sites $x \in L_i$ with lattice sites $y \in L_j$.

Restriction and Prolongation The main ingredient for restriction R and prolongation P, cf. (15), is an approximate inverse of A_{ff}. Since the matrix A_{ff} is γ_5-symmetric, we consider only approximate inverses which reflect the γ_5-symmetry,

$$\tilde{A}_{ff}^{-1} = \Gamma_5^{(f)} \tilde{A}_{ff}^{-H} \Gamma_5^{(f)}. \tag{22}$$

At a glance, it is not necessary to use the same approximate inverse for restriction and prolongation, but it seems to be a quite natural choice which ensures in conjunction with (22) that restriction R and prolongation P are γ_5-adjoint,

$$\Gamma_5 R^H = P \Gamma_5^{(c)}. \tag{23}$$

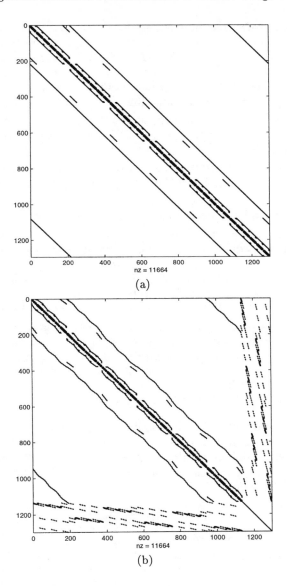

Fig. 2. Graph of Wilson fermion matrix with a lexicographic ordering (a) and a fine coarse ordering (b) of the lattice sites.

Since $A_{ff} = I - \kappa D_{ff}$, the inverse of A_{ff} formally is given by

$$A_{ff}^{-1} = \sum_{\nu=0}^{\infty} (\kappa D_{ff})^{\nu}. \tag{24}$$

A truncation of the Neumann series yields, e.g., $\tilde{A}_{ff}^{-1} = I$ or

$$\tilde{A}_{ff}^{-1} = I + \kappa D_{ff}. \tag{25}$$

For the trivial case, $\tilde{A}_{ff}^{-1} = I$, we obtain both a nearest-neighbour restriction R and a nearest-neighbour prolongation R,

$$R = \left(\kappa D_{cf} \; I\right), \quad P = \begin{pmatrix} \kappa D_{fc} \\ I \end{pmatrix}, \tag{26}$$

whereas in the non-trivial case (25), restriction R and prolongation P also take next-to-nearest neighbours, i.e. L_2 sites, into consideration,

$$R = \left(\kappa D_{cf}(I + \kappa D_{ff}) \; I\right), \quad P = \begin{pmatrix} \kappa(I + \kappa D_{ff})D_{fc} \\ I \end{pmatrix}. \tag{27}$$

An incomplete LU (and SGS) decomposition leads to an approximate inverse,

$$\tilde{A}_{ff}^{-1} = \begin{pmatrix} I & 0 & \kappa D_{L_1,L_2} \\ 0 & I & \kappa D_{L_3,L_2} \\ 0 & 0 & I \end{pmatrix} \begin{pmatrix} I & 0 & 0 \\ 0 & I & 0 \\ \kappa D_{L_2,L_1} & \kappa D_{L_2,L_3} & I \end{pmatrix} \tag{28}$$

which is often proposed in literature. For the Wilson fermion matrix, using the incomplete LU preconditioner of A_{ff} as an approximate inverse within the restriction R and prolongation P is identical to the usage of the truncated Neumann series

$$\tilde{A}_{ff}^{-1} = I + \kappa D_{ff} + (\kappa D_{ff})^2 \tag{29}$$

since all information collected on L_2 sites is neglected by D_{cf}.

For a cold configuration, i.e. $U_\mu(x) = I$, we suggest to use a nearest and next-to-nearest neighbour restriction and prolongation (27), where the diagonal and the off-diagonal part of the approximate inverse are supplemented with additional weights

$$\tilde{A}_{ff}^{-1} = \alpha_0 I + \alpha_1 \kappa D_{ff}. \tag{30}$$

The weights can be adjusted such that the prolongation P maps constant gridfunctions onto constant ones,

$$\alpha_0 = \frac{1}{2\kappa}, \quad \alpha_1 = \frac{1}{16\kappa^2}. \tag{31}$$

This reveals the relation to multigrid solvers with geometric restriction and prolongation operators [14].

In the general case the choice (31) makes no sense, and we propose to use either a few steps of a basic iteration scheme to solve systems with A_{ff} approximately or (30) as an approximate inverse, where the parameters are choosen such that the term $\alpha_0(1 + \alpha_1 t)(1 - t)$ is minimal over the spectrum of D_{ff} in a given norm.

Coarse Grid Matrix The sparsity of the coarse grid matrix is crucial for the efficiency of the two-level preconditioner. To obtain a recursive procedure, we prescribe the sparsity pattern to avoid fill-in and seek for an approximate inverse such that the Schur complement approximation still represents a local coupling.

Due to the selection of the coarse grid Ω_c, for the exact Schur complement it is clear that coarse sites are only able to interact with coarse sites via link paths of even length,

$$S_A = I - \kappa^2 D_{cf} \sum_{\nu=0}^{\infty} (\kappa D_{ff})^{\nu} D_{fc} = I - \kappa^2 D_{cf} \sum_{\nu=0}^{\infty} (\kappa D_{ff})^{2\nu} D_{fc}. \qquad (32)$$

Thus, if we truncate the Neumann series in (32) such that we obtain a nearest neighbour coupling as an approximate inverse,

$$\widetilde{A}_{ff}^{-1} = I + \kappa D_{ff}, \qquad (33)$$

the off-diagonal part κD_{ff} does not contribute to the Schur complement, cf. (32). Hence, (33) leads immediately to a Schur complement approximation $\widetilde{S}_A$,

$$\widetilde{S}_A = I - \kappa^2 D_{cf} D_{fc}. \qquad (34)$$

Like the original Wilson fermion matrix A, the coarse grid matrix $\widetilde{S}_A$ describes a nearest neighbour coupling on the coarse sites Ω_c,

$$\psi_c(x) - 2\kappa^2 \left(\sum_{\mu=1}^{4} \left((I_4 - \gamma_\mu) \otimes U_\mu(x) U_\mu(x + e_\mu) \right) \psi_c(x + 2e_\mu) \right. \qquad (35)$$

$$\left. + \sum_{\mu=1}^{4} \left((I_4 + \gamma_\mu) \otimes U_\mu^H(x - e_\mu) U_\mu^H(x - 2e_\mu) \right) \psi_c(x - 2e_\mu) \right) = \varphi_c(x).$$

Actually, there are two independent coarse grid systems corresponding to L_0 and L_4 sites, respectively. All basic properties of the Wilson fermion matrix mentioned in Sect. 2 are maintained on the coarse grid, i.e. the coarse grid matrices are γ_5-symmetric and exhibit the odd-even symmetry. In this context, the coarse grid matrices can be regarded as new Wilson fermion matrices with

$$\kappa' = 2\kappa^2, \quad U'_\mu(x) = U_\mu(x) U_\mu(x + e_\mu). \qquad (36)$$

Suppose that both restriction and prolongation are optimal,

$$R = \left(-A_{cf} A_{ff}^{-1} \; I \right), \quad P = \begin{pmatrix} -A_{ff}^{-1} A_{fc} \\ I \end{pmatrix}, \qquad (37)$$

a simple but tedious calculation yields

$$M^{-1}A = \begin{pmatrix} I & -A_{ff}^{-1}A_{fc} \\ 0 & I \end{pmatrix} \begin{pmatrix} I & 0 \\ 0 & \widetilde{S}_A^{-1}S_A \end{pmatrix} \begin{pmatrix} I & A_{ff}^{-1}A_{fc} \\ 0 & I \end{pmatrix}. \tag{38}$$

Since the lowest eigenmode of $\widetilde{S}_A$ should be an approximation of the lowest eigenmode of S_A, we suggest to introduce (at least) an additional weight ω to control the influence of the hopping terms on the coarse lattice,

$$\widetilde{S}_A = I - \omega\kappa^2 D_{cf}D_{fc} \tag{39}$$

For a cold configuration the lowest eigenmode $(\lambda_{\min}(S_A), \mathbf{1})$ can be approximated exactly,

$$\widetilde{S}_A\mathbf{1} = \lambda_{\min}(S_A)\mathbf{1} \iff \widetilde{S}_A = I - \omega\kappa^2 D_{cf}D_{fc}, \quad \omega = \frac{1}{1-48\kappa^2}. \tag{40}$$

For a realistic or, even worse, a hot configuration, a scalar ω seems not to be sufficient. For the latter case, we propose to use a (block) diagonal matrix.

Although, a cold (hot) configuration $U_\mu(x)$ remains a cold (hot) configuration $U_\mu'(x)$ on the coarse lattice, for realistic configurations the resulting coarse grid configurations become more random, see Fig. 3.

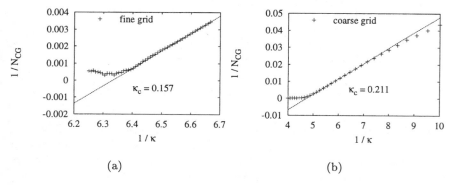

(a)   (b)

Fig. 3. Fit of critical hopping parameter κ_c for a realistic configuration (a) and the resulting coarse grid configuration (b). Here, the number of cg iterations to solve the normal equation is taken as a measure for the condition number.

Thus, even from a physical point of view, the introduction of additional weights may be motivated to preserve physical properties. In this case, the weights should be adapted carefully subject to the given lattice spacing, the physical coupling β and the quark mass m_q.

Smoothing Unfortunately, for a combination of multi-level preconditioning with smoothing procedures as suggested in [15], it is not possible to prove the smoothing property for any simple smoothing procedure in the case of the Wilson fermion matrix. For this reason, we change over to the indefinite, hermitian Wilson fermion matrix $Q = \Gamma_5 A$ and remark that its exact factorization yields immediately a two-level preconditioner for Q,

$$M_Q^{-1} = \begin{pmatrix} I & -\tilde{A}_{ff}^{-1}A_{fc} \\ 0 & I \end{pmatrix} \begin{pmatrix} \tilde{A}_{ff}^{-1}\Gamma_5^{(f)} & 0 \\ 0 & \tilde{S}_A^{-1}\Gamma_5^{(c)} \end{pmatrix} \begin{pmatrix} I & 0 \\ -\Gamma_5^{(c)}A_{cf}\tilde{A}_{ff}^{-1}\Gamma_5^{(f)} & I \end{pmatrix}, \quad (41)$$

where the approximate inverse of A_{ff} and the Schur complement approximation are choosen as explained above. Now, the restriction R_Q and prolongation P_Q writes

$$R_Q = \begin{pmatrix} -\Gamma_5^{(c)}A_{cf}\tilde{A}_{ff}^{-1}\Gamma_5^{(f)} & I \end{pmatrix}, \quad P_Q = \begin{pmatrix} -\tilde{A}_{ff}^{-1}A_{fc} \\ I \end{pmatrix}, \quad (42)$$

respectively. If $\tilde{A}_{ff}^{-1}$ is γ_5-symmetric, then $R_Q = P_Q^H$ holds.

For the hermitian Wilson fermion matrix Q it is easy to verify that the modified Jacobi smoother [14],

$$S = (I - \theta Q)(I + \theta Q), \quad \theta = 1/\|Q\|_2, \quad (43)$$

fulfills

$$\|QS^\nu\|_2 \le \eta_1(\nu)\|Q\|_2, \quad \eta_1(\nu) = \frac{1}{\sqrt{2\nu+1}}\left(\frac{2\nu}{2\nu+1}\right)^\nu. \quad (44)$$

More generally, the application of ν smoothing steps with smoother S,

$$S^\nu = s(\theta Q), \quad s(t) = 1 - t^2 q(t^2), \quad \theta = 1/\|Q\|_2, \quad (45)$$

implies

$$\|QS^\nu\|_2 = \frac{1}{\theta}\|\theta Q s(\theta Q)\|_2 = \|p(\theta Q)\|_2\|Q\|_2, \quad p(t) = ts(t). \quad (46)$$

The problem of minimizing

$$\max\{|p(t)| : |t| \le 1\} \quad (47)$$

over all polynomials p, $p(t) = ts(t)$, $\deg(p) \le 2m$, can be solved by

$$p(t) = \frac{(-1)^m}{2m+1}T_{2m+1}(t), \quad |p(t)| \le \frac{1}{2m+1} = \eta_2(\nu), \quad (48)$$

where T_m are the well-known Chebyshev polynomials,

$$T_m(t) = \cos(m \arccos(t)), \quad \deg(T_m) = m. \quad (49)$$

This approach, which is is strongly related to conjugate gradient smoothing, is known as semi-iterative smoothing [14]. To compare the modified Jacobi smoother and the semi-iterative smoothing procedure, in Table 1 smoothing numbers $\eta(\nu)$ are given. Although the semi-iterative smoother is expensive, now an optimal smoothing procedure for the hermitian Wilson fermion matrix is available.

Table 1. Smoothing numbers $\eta(\nu)$ for modified Jacobi and semi-iterative smoother

ν	1	2	3	4	5	6	7	8
$\eta_1(\nu)$	0.385	0.286	0.238	0.208	0.187	0.176	0.159	0.149
$\eta_2(\nu)$	0.333	0.2	0.143	0.111	0.091	0.077	0.067	0.059

4   Numerical Results

As a simple *model problem*, we consider a nearest-neighbour coupling on a two-dimensional regular 16×16 lattice with periodic boundary conditions. The generic equation $A\psi = \varphi$ writes

$$\psi_x - \kappa \sum_{\mu=1}^{2} \left(U_\mu(x)\psi_{x+e_\mu} + U_\mu^H(x - e_\mu)\psi_{x-e_\mu}\right) = \varphi_x, \tag{50}$$

where the matrices $U_\mu(x) \in U(1)$ are either constant (cold) or randomly (hot) generated, and the parameter κ corresponds to $m_q = 0.01$.

First, we consider the choice of an appropriate approximate inverse of A_{ff}. Since $\rho(D_{ff}) \leq \rho(|D_{ff}|) = \sqrt{8}$, the spectrum of A_{ff} can be bounded,

$$0 \ll 1 - \sqrt{8}\kappa \leq \lambda(A_{ff}) \leq 1 + \sqrt{8}\kappa, \tag{51}$$

which implies that the matrix A_{ff} is well-conditioned,

$$\kappa(A_{ff}) = \frac{1 + \sqrt{8}\kappa}{1 - \sqrt{8}\kappa} < \frac{1 + \sqrt{8}\kappa_c}{1 - \sqrt{8}\kappa_c}. \tag{52}$$

Hence, the matrix A_{ff} can be approximated easily. For instance, for $\widetilde{A}_{ff}^{-1} = I + \kappa D_{ff}$ we get

$$\kappa(\widetilde{A}_{ff}^{-1} A_{ff}) = \kappa(I - \kappa^2 D_{ff}^2) \leq \frac{1}{1 - 8\kappa^2} < \frac{1}{1 - 8\kappa_c^2}. \tag{53}$$

Now, for a cold configuration we set the approximate inverse such that the prolongation maps constant gridfunctions onto constant ones, cf. (31), and as Schur complement approximation we consider (40). In contrast to the cold case, for a hot configuration we simply use the truncated Neumann series (25) as approximate inverse and (39) as Schur complement approximation.

In Fig. 4 the spectra of unpreconditioned and preconditioned A_{ff} are plotted for our model problem. As expected, the eigenvalues of the preconditioned matrix are clustered well which indicates that in both cold and hot case the approximate inverse is suitable.

Table 2. Condition numbers $\kappa(A) = \lambda_{\max}(A)/\lambda_{\min}(A)$.

configuration	$\kappa(A)$	$\kappa(A_e)$	$\kappa(A_{ff})$	$\kappa(\widetilde{A}_{ff}^{-1}A_{ff})$	$\kappa(\widetilde{S}_A^{-1}S_A)$	$\kappa(M^{-1}A)$
cold	401.0000	100.7506	5.7474	2.0041	1.6920	3.3125
hot	347.5989	87.3956	8.2013	2.5808	19.9083	45.0318

Figure 5 shows the spectra of the Schur complement and its approximation as well as the eigenvalue distribution of the preconditioned Schur complement. For the cold configuration, the lowest eigenvalue of S_A is approximated exactly which results in an almost optimal clustering of the eigenvalues of the preconditioned Schur complement. For a hot configuration, the situation is worse. Nevertheless, we gain an improvement of a factor of about 7.7 in comparison to the original matrix A and still a factor of about 2 in comparison to the odd even reduced system with a simple Schur complement approximation, cf. Tab. 2. We believe that with a more sophisticated nearest neighbour coupling as Schur complement approximation the situation is more favorable. For a realistic configuration, we expect to end up with condition numbers somewhere between the condition numbers of a cold and a hot configuration.

5   Conclusions

We have presented a decomposition of the lattice sites such that the coarse sites can be interpreted as two standard coarse grids interlaced into the whole lattice. The resulting first sub-matrix, which governs the coupling of fine sites with fine ones and which has to be approximated to obtain practical preconditioners, is well-conditioned. Hence, a truncated Neumann series as approximate inverse is sufficient. Alternatively, a few steps of a basic iteration scheme might be performed. Due to the selection of the coarse sites, we have shown how to achieve a Schur complement approximation which can be perceived as two Wilson fermion matrices. An extension of the two-level preconditioner to the multi-level case is straightforward. Motivated by results of a simple model problem, the described approach seems to be suitable for Wilson fermions, and first experiments in this area are promising.

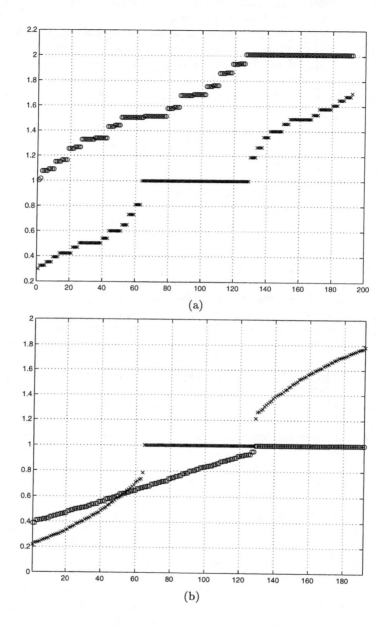

Fig. 4. Eigenvalue distribution of A_{ff} ($\times$) and $\widetilde{A}_{ff}^{-1}A_{ff}$ (o) for a cold (a) and a hot (b) configuration.

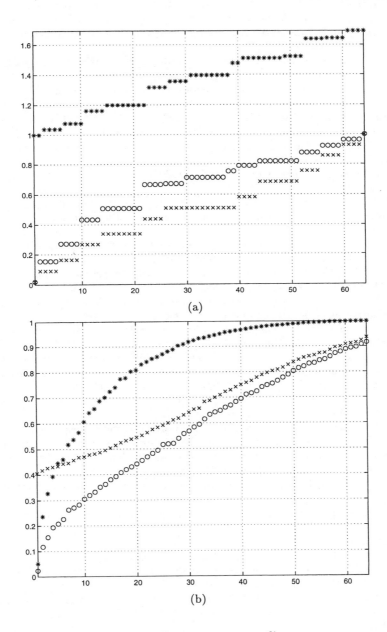

Fig. 5. Eigenvalue distribution of $\widetilde{S}_A$ ($\times$), S_A (o) and $\widetilde{S}_A^{-1} S_A$ ($*$) for a cold (a) and a hot (b) configuration.

References

1. O. Axelsson, P. Vassilevski: Algebraic multilevel preconditioning methods – Part I, Numer. Math. 56, 157–177 (1989)
2. O. Axelsson, P. Vassilevski: Algebraic multilevel preconditioning methods – Part II, SIAM J. Num. Anal. 27, 1569–1590 (1990)
3. R. Barrett, M. Berry, T. Chan, J. Demmel, J. Donato, J. Dongarra, V. Eijkhout, R. Pozo, C. Romine, H. van der Vorst: Templates for the Solution of Linear Systems: Building Blocks for Iterative Methods, SIAM, Philadelphia (1994)
4. W. Bietenholz, N. Eicker, A. Frommer, Th. Lippert, B. Medeke, K. Schilling, G. Weuffen, Preconditioning of Improved and "Perfect" Fermion Actions, Comput.Phys.Commun. 119 (1999) 1-18
5. P. de Forcrand: Progress on lattice QCD algorithms, Nucl.Phys.Proc.Suppl. 47 (1996) 228-235
6. N. Eicker, W. Bietenholz, A. Frommer, Th. Lippert, B. Medeke, K. Schilling, A Preconditioner for Improved Fermion Actions, Nucl.Phys.Proc.Suppl. 73 (1999) 850-852
7. S. Fischer, A. Frommer, U. Glässner, Th. Lippert, K. Schilling: A Parallel SSOR Preconditioner for Lattice QCD, Comp. Phys. Comm. 98, 20–34 (1996)
8. R. Fletcher, Conjugate gradient methods for indefinite systems, in: Lecture Notes in Mathematics 506, Springer Verlag, 1976
9. R. Freund, N. Nachtigal: QMR: a quasi-minimal residual method for non-Hermitian linear systems, Numer. Math. 60, 315–339 (1991)
10. A. Frommer: Linear systems solvers - recent developments and implications for lattice computations, Nuclear Physics B, 53 (Proc. Suppl.) 120-126 (1997)
11. A. Frommer, Th. Lippert, B. Medeke, K.Schilling (edts.). Numerical Challenges in Lattice Quantum Chromodynamics. Proceedings of the Interdisciplinary Workshop on Numerical Challenges in Lattice QCD, Wuppertal, August 22-24, 1999. Series Lecture Notes in Computational Science and Engineering (LNCSE). Springer Verlag, Heidelberg, 2000.
12. A. Frommer, B. Medeke: Exploiting Structure in Krylov Subspace Methods for the Wilson Fermion Matrix, in: 15th IMACS World Congress on Scientific Computation, Modelling and Applied Mathematics (A. Sydow, ed.), Wissenschaft & Technik Verlag, Berlin, 485–490 (1997)
13. Gutknecht, M. H.: Remarks on Lanczos-type methods for Wilson fermions, in [11]
14. W. Hackbusch: Multi-grid methods and applications, Springer, Heidelberg (1985)
15. Notay, Y.: Optimal V-cycle algebraic multilevel preconditioning. (1997)
16. Notay, Y.: On Algebraic Multilevel Preconditioning, in [11]
17. Reusken, A.: An Algebraic Multilevel Preconditioner for Symmetric Positive Definite and Indefinite Systems, in [11]
18. Y. Saad, M. Schultz: GMRES: a generalized minimal residual algorithm for solving nonsymmetric linear systems, SIAM J. Sci. Stat. Comp. 7, 856–869 (1986)
19. H. van der Vorst: Bi-CGStab: A fast and smoothly converging variant of Bi-CG for the solution of nonsymmetric linear systems, SIAM J. Sci. Stat. Comp. 13, 631–644 (1992)

Stochastic Estimator Techniques for Disconnected Diagrams

Stephan Güsken

Bergische Universität Wuppertal
Fachbereich Physik
42097 Wuppertal, Germany

Abstract. The calculation of physical quantities by lattice QCD simulations requires in some important cases the determination of the inverse of a very large matrix. In this article we describe how stochastic estimator methods can be applied to this problem, and how such techniques can be efficiently implemented.

1 Introduction

Within our current level of comprehension of the fundamental principles of nature, physical processes on an atomic or subatomic scale can be successfully described by Quantum Field Theories (QFT). In such theories particles as well as their interactions are represented by quantum fields, defined at each space-time point. The value of a physical quantity, which can be measured in experiment, can be calculated by a weighted average over all "would be" values of this quantity, achieved for each possible configuration of the quantum fields involved. The weight with which each of these "would be" values contributes is determined by the so-called action, a scalar quantity which contains the characteristic features of the QFT in question and which depends on the quantum fields. The formal expression of this averaging procedure is known as the path functional.

An exact analytical treatment of the path functional is in most cases not possible. Approximate solutions can be achieved in the framework of perturbation theory if the interaction strength is weak. Perturbative methods have been proven very successful in the evaluation of the QFT of the electromagnetic and weak forces. They fail however when applied to the QFT of the strong force, the so called Quantum Chromo Dynamics (QCD). The strong force is responsible for a large range of phenomena at and below the scale of the atomic nuclei.

Lattice QCD is designed for a non-perturbative numerical evaluation of QCD. The space-time continuum is approximated by a lattice with $N_s^3 \times N_t$ space-time points. The calculation of physical quantities is done in two steps. First, one generates a representative sample of quantum field configurations, where each configuration is drawn according to its specific weight, by a Monte

Carlo procedure [1–3]. Secondly, one determines the "would be" value of the physical quantity in question on each of the quantum field configurations and takes the average. We call the latter step the analysis of quantum field configurations.

Clearly, the computational effort which has to be invested in the analysis part depends on the physical quantity one is interested in. In this article we report on a computationally very hard problem which occurs in the analysis of configurations with respect to so called disconnected contributions. The latter are of great physical importance as they are expected to play a substantial rôle in the solution of the "proton spin problem" [4] and of the "$U(1)$ problem" of QCD [5].

Naively, the calculation of disconnected contributions requires the inversion of a complex matrix of size $(N_s^3 \times N_t \times 12)^2$ on each single quantum field configuration. For currently available lattice sizes, see ref.[6], such a calculation would be prohibitively expensive. To circumvent this problem one applies stochastic estimator methods which converge to the true result in the stochastic limit. It turns out however, that even with such techniques one still needs a parallel supercomputer to handle this problem.

This article is organized as follows. In the next section we will give an impression of the physical meaning of disconnected contributions. Section 3 explains what has to be done technically to calculate such contributions. Section 4 is devoted to the stochastic estimator techniques. The implementation of these techniques on parallel computers is discussed in section 5. Section 6 will give an overview of state of the art calculations of disconnected contributions.

2   The Physical Motivation

The "parton" picture of a proton as being made of 3 interacting quarks is not applicable in full QCD. The reason is that in this case the spontaneous creation and annihilation of quark antiquark pairs leads to an additional contribution to the proton amplitude.

Suppose that we would like to investigate the properties of a proton by a scattering experiment, e.g. by deep inelastic μ-p (muon proton) scattering. Then, the μ-p scattering amplitudes measured in such an experiment could in principle differ sizeably from the parton expectation since the latter neglects the interaction of the μ particle with a quark-antiquark loop.

To illustrate this point we show in fig. 1 the full QCD contributions to the propagator of a proton which interacts with an external current j.   Part (a) of the figure depicts the naive (parton) case, where the current couples to one of the quarks of the proton. Part (b) shows the interaction of the current j with a quark-antiquark loop, in the field of the proton. This disconnected contribution is present only in full QCD. We emphasize that the location in space and time of the quark antiquark loop is not fixed. Thus, to calculate the disconnected contribution one has to sum over all possible positions.

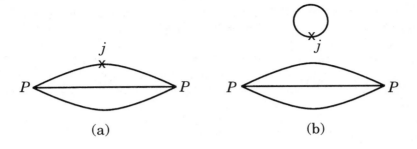

Fig. 1. *Connected (a) and disconnected (b) contributions to a proton interacting with a current j. All quark lines, including the quark loop, are connected by infinitely many gluon lines and virtual quark loops.*

Of course, it is not obvious from these considerations that the disconnected part really yields non-negligible contributions to the scattering amplitudes. In fact, it turns out that many of them have a structure such that their net contribution is expected to be small.

There is however a class of amplitudes, the flavor singlet amplitudes, where disconnected contributions can be sizeable. In order to investigate quark loop effects in QCD it is therefore of utmost interest to calculate flavor singlet amplitudes and to compare the results with experimentally measured data.

From experimental measurements one can extract the values of at least 2 flavor singlet amplitudes. The first, which describes the interaction of a proton with a pseudo-vector current deviates by about a factor of 2 from the naively expected value. This deviation gave rise to the so called "proton spin crisis". The second, which couples a scalar current to a proton, yields, when multiplied by the quark mass, the pion-nucleon sigma term $\Sigma_{\pi N}$ [8]. The experimental value of this quantity also differs by about a factor of 2 from the naive expectation. Thus, these quantities are most promising candidates to study the influence of quark loops by a full QCD lattice simulation.

Disconnected amplitudes are supposed to contribute also to many physical processes other than proton scattering. For example, a (pseudo scalar) meson, which is made of a quark and an antiquark, can be "mimicked", with respect to its quantum numbers, by 2 quark antiquark loops. This is shown in fig.2. An experimental measurement of e.g. the mass of this meson would include both terms, connected and disconnected.

From symmetry considerations one again concludes, that the disconnected part should contribute mostly if the quarks in the meson are put together in a flavor singlet combination. Experimentally, one finds that the mass of such a flavor singlet meson, which is named η', is much larger than that of its

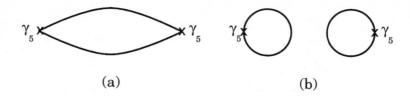

Fig. 2. *Connected (a) and disconnected (b) contributions to the propagator of a pseudo scalar meson.*

non-singlet partners. This discrepancy is called the "$U(1)$ problem of QCD".

Clearly, a full QCD lattice calculation of the diagrams of fig. 2 would be of great help to solve this problem.

3   The Technical Problem

3.1   Quark Propagator

A key quantity in the analysis of quantum field configurations is the quark propagator Δ. It is defined as the correlation of 2 (fermionic) quantum fields Ψ and $\bar{\Psi}$ at the space-time points (x, t) and (x', t') respectively:

$$\Delta(x, t, a, \alpha; x', t', a', \alpha') = \psi(x, t, a, \alpha)\bar{\psi}(x', t', a', \alpha') . \tag{1}$$

The indices $a, \alpha,\ a', \alpha'$ denote internal degrees of freedom of the fermionic quantum fields. a, a' are called color indices. They can take the values 1,2 and 3. α, α' are called Dirac indices. They run from 1 to 4.

In the language of QCD, the quark propagator Δ denotes the probability amplitude of a strongly interacting elementary particle (quark) to travel from point (x, t) to point (x', t'). Once Δ is known, a whole bunch a physical quantities like the spectrum and the decay properties of strongly interacting composite particles can be determined immediately.

Unfortunately, eq. (1) cannot be used for a numerical calculation of Δ since, with current Monte Carlo algorithms, the fermion fields $\Psi, \bar{\Psi}$ are not explicitly available. They enter only indirectly in form of the fermionic matrix M. The quark propagator is related to M by

$$\Delta(x; x') = M^{-1}(x, x') , \tag{2}$$

where we have used the multi index x for space-time, color and Dirac indices.

The fermionic matrix M is complex and sparse. In the widely used Wilson form it is given by

$$M(x, a, \alpha; x', a', \alpha') = \tag{3}$$

$$1 - \kappa \sum_{\mu=1}^{4} [(1 - \gamma_\mu(\alpha; \alpha'))U_\mu(a; a')(x)\delta_{x+\hat{\mu};x'} +$$

$$(1 + \gamma_\mu(\alpha; \alpha'))U_\mu^\dagger(a; a')(x - \hat{\mu})\delta_{x-\hat{\mu};x'}] \ .$$

Here, κ is a real number, which determines the mass of the propagating quark. γ_μ denotes the 4×4 anti-commuting Dirac matrices. The "links" U_μ are $SU(3)$ matrices, which act in color space. They represent the quantum fields of the interaction between quarks. The unit vector $\hat{\mu}$ points into the direction μ.

According to eq. (2), the computationally expensive part of the analysis is to determine the inverse of M for each quantum field configuration U. Fortunately, for many applications, it is not necessary to solve the full problem. For example, to calculate the spectrum and the decay properties of strongly interacting composite particles, it is sufficient to determine only one row of M^{-1}. This reduced problem

$$M(z, x)\Delta(x, x_0) = \delta_{z, x_0} \quad , \quad x_0 \text{ fixed} \tag{4}$$

can be treated using fast iterative solvers [9] with moderate computational effort.

3.2   Disconnected Contributions

There is however a class of important physical quantities (see above), whose determination requires, in a sense, the solution of the full problem. To be specific, the prominent combinations D_Γ of M^{-1} needed for the calculation of the disconnected contributions are given by

$$D_\Gamma = Tr\left[\Gamma M^{-1}\right] \ , \tag{5}$$

where $\Gamma = 1, \gamma_\mu, \gamma_5, \gamma_\mu\gamma_5$ is a 4×4 matrix which acts on the Dirac indices of M^{-1}. γ_5 is defined by $\gamma_5 = \gamma_1\gamma_2\gamma_3\gamma_4$.

Clearly, an exact determination of D_Γ would require $N_s^3 \times N_t \times 3 \times 4$ applications of the "row" method, eq. (4). This would overtax even the capacity of a fast parallel supercomputer.

We will see in the next section how one can circumvent this problem by a calculation of a reliable estimate, D_Γ^E, instead of the exact value D_Γ.

4   Stochastic Estimation

Suppose that we would have created N random vectors $\eta_k(i)$, $i = 1, \ldots, V$, $k = 1, \ldots, N$ with the properties

$$\lim_{N\to\infty} \frac{1}{N} \sum_{k=1}^{N} \eta_k(i) \equiv \langle \eta(i) \rangle = 0 \,, \tag{6}$$

$$\lim_{N\to\infty} \frac{1}{N} \sum_{k=1}^{N} \eta_k(i)\eta_k(j) \equiv \langle \eta(i)\eta(j) \rangle = \delta_{i,j} \,. \tag{7}$$

These properties are fulfilled by, for example, Gaussian [10] or Z_2 [11,12] random number distributions.

Suppose furthermore that we would modify eq. (4) by inserting a random source vector η_k on the right hand side

$$M(z,x)\tilde{\Delta}_k(x) = \eta_k(z) \,, \quad \tilde{\Delta}_k(x) = \delta_{x,x'}\eta_k(x') \,. \tag{8}$$

Then, the product $\eta_k \tilde{\Delta}$ can be written as

$$\tilde{D}_1^k = Tr(\Delta) \sum_{i=1}^{V} \eta(i)_k \eta(i)_k + \sum_{i \neq j} \Delta(i,j)\eta(i)_k\eta(j)_k \,. \tag{9}$$

According to eq. (7) one gets in the stochastic limit ($N \to \infty$) of N solutions to eq. (8)

$$\langle \tilde{D}_1 \rangle = D_1 \,. \tag{10}$$

Thus, this procedure converges to the correct result. We mention that the stochastic estimator method can be also applied, with small modifications [13], to the calculation of arbitrary D_Γ, c.f. eq. (5).

Of course, the stochastic method is useful only if already a moderate number N of solutions to eq. (8) suffices to calculate a reliable estimate

$$D_1^E = \frac{1}{N} \sum_{k=1}^{N} \tilde{D}_1^k \tag{11}$$

of D_1. The question of how large N should be chosen has been investigated by the authors of ref. [14] in some detail for a medium size lattice ($V = 16^3 \times 32 \times 3 \times 4$). It turned out that $N \simeq 100$ allows to estimates D_1 within a 10% uncertainty. The situation might be much less favorable however for $\Gamma \neq 1$. For example, for

$$D_\Gamma = D_{\gamma_3\gamma_5} = \tag{12}$$

$$\sum_{x,a} [\, \Delta(x,a,1;x,a,1) - \Delta(x,a,2;x,a,2)$$

$$-\Delta(x,a,3;x,a,3) + \Delta(x,a,4;x,a,4)\,]$$

one has to determine differences of the diagonals of Δ instead of the sum over diagonal elements. Since all these numbers are of similar size, such a task could require a much higher number of estimates to achieve a reliable result on each single quantum field configuration.

Fortunately, the problem is softened by the average over quantum field configurations, for the following reason: Quantum Field Theories, such as QCD, exhibit the property of gauge invariance, i.e. physical quantities do not change their values under gauge transformations. The path integral, which represents the formal expression of how to calculate physical quantities in QFT, automatically removes all non gauge invariant contributions. Since most of the (unwanted) noise terms on the right hand side of eq. (9) are not gauge invariant, the average over quantum field configurations will help to increase the accuracy of the estimate of D_Γ^E. Nevertheless, as we will show at the end of this article, one still needs at least 400 estimates per quantum field configuration to achieve statistically significant signals for D_Γ or for the (physically important) correlations between D_Γ and the proton propagator.

Thus, the computational effort which is necessary to calculate disconnected contributions exceeds the one for the "standard analysis" by more than 2 orders of magnitude.

5   Parallelization

There are a least two straightforward ways to implement the numerical problem defined by eqs. (8),(11) on a parallel computer. The first, a farming approach, can be used for medium size lattices on machines with a, compared to the processor speed, slow communication network. The second one, a non-trivial parallelization, is useful for large lattice and, on machines with fast communication lines.

5.1   Farming

A natural way to implement the stochastic estimator method on a parallel computer arises from the fact that the estimates, eq. (8), are completely independent of each other. Thus, the estimates can be calculated simultaneously on separate compute nodes. Communication is required only at the beginning of the calculation, when quantum fields and stochastic sources have to be passed to their respective nodes, and at the end, when the results of the single estimates have to be gathered and averaged.

Alternatively, one can implement the farming idea with respect to the quantum field configurations. In this case, each processor receives its own quantum field configuration at the beginning and computes N estimates.

In both cases, one has to ensure that the random numbers used on such a distributed system are not correlated. This can be achieved either by running a large period random number generator only on one node, which passes the random vectors successively to all other nodes, or by using a parallel random number generator, where each node creates its own random numbers from an independent stream of the generator [15].

The ideal machine to implement a stochastic estimator program in farming style is a large cluster of powerful workstations, which are connected e.g. by Ethernet.

Let us give an example. One needs with a standard inverter, say the minimum residual (MR) inverter, on a workstation which runs with a sustained speed of 50 Mflops about 20 minutes of CPU time to solve eq. (8) on a $16^3 \times 32$ lattice for values of κ, c.f. eq. (3), in the physically interesting range. Thus, on a cluster of 100 workstations it would take about 11 days to calculate D_Γ^E with 400 estimates on 200 quantum field configurations.

The memory requirements on each node for medium size lattices are moderate. For a $16^3 \times 32$ lattice one needs an overall amount of about 60 Mbytes. Thus, even a $26^3 \times 52$ lattice would easily fit into the memory of a 512 Mbyte workstation.

5.2   Decomposition

To handle large lattices on parallel computers with a comparatively small amount of memory per node or on massively parallel systems with a large number of nodes, one should divide the lattice into sub-lattices and distribute the latter among the nodes. Since the matrix M, see eq. (3), connects only nearest neighbors, this can be accomplished in a straightforward way. A simple but efficient realization of this decomposition approach is shown in fig. 3. for an 8×8 lattice. Each node administrates the data points of a 4×4 sub-lattice (denoted by crosses) as well as the current values of the surface points of the neighboring sub-lattices (denoted by O). After each iteration step of the solver, e.g. MR, the updated values of each sub-lattice (X_N, X_S, X_W, X_E) are passed to the "O-buffers" of the neighboring sub-lattices, i.e. $X_S \to O_N, X_N \to O_S$ etc. . This procedure is repeated until some stopping criterion, set by e.g. an upper bound of the norm of the rest vector, is fulfilled.

Decomposition requires of course a communication network of much higher quality than the farming approach. Although the amount of data which has to be exchanged between the nodes after each iteration step can be adjusted by a suitable choice of the sub-lattice size, the number of communications necessary to complete one full estimate can not: It is determined by the number of iterations needed by the solver to converge. Typically, this number is in the range of a few hundred for κ values in a physically interesting region. Thus, the startup time for the communication has to be taken seriously into account.

We mention that the memory overhead introduced by the "O-buffers" can be avoided on shared memory machines. On such a machine each processor reads the required data directly from the memory of the neighboring nodes.

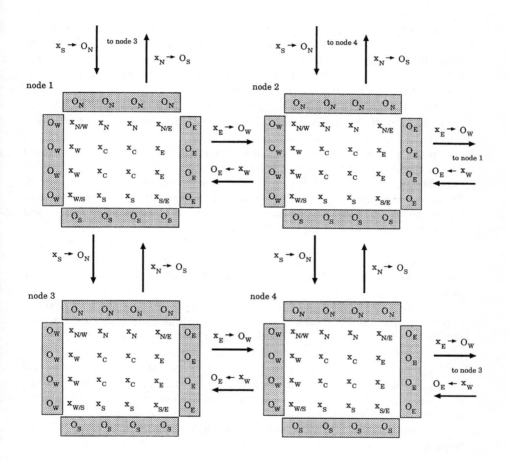

Fig. 3. *Internal parallelization of a 8 × 8 lattice with periodic boundary conditions lattice on 4 node system.*

6 Summary

The evaluation of disconnected contributions with stochastic estimator methods is still at its beginning. Clearly, this is due to the requirement of a very high computer speed needed to tackle such a problem.

Pioneering studies of disconnected contributions have been performed some time ago on conventional (vector) computers in the quenched approximation of QCD, where one neglects the fermionic quantum fields in the Monte Carlo update, by the authors of refs. [16,17] and [18,19]. Due to the lack of computer speed, the authors had to make concessions to the statistical reliability of their results. Thus, although the data look promising, a systematic bias of their findings cannot be excluded.

With the advent of powerful parallel computers it became possible to treat disconnected contributions more reliably [13,14,20,21], although one is still limited to medium size lattices.

So far, the most intense study of such contributions has been performed by the SESAM collaboration [13,20] on a QH2 APE-100 computer [22] which runs the stochastic estimator code with a sustained speed of $\simeq 6$ Gflops. SESAM has analyzed 200 full QCD quantum field configurations of a $16^3 \times 32$ lattice, at several values of the mass parameter κ. On each configuration, and for each κ, the values of D_1 and $D_{\gamma_3\gamma_5}$ have been estimated 400 times.

SESAM uses Z_2-noise techniques[14] as well as a plateau accumulation method (PAM) for noise/signal improvements[13]. Further details and novel approaches concerning improvements can be found in the contribution by Walter Wilcox[23].

The statistical analysis of the SESAM data revealed that the related physical quantities, i.e. the correlation C_Γ of D_Γ and the proton propagator with respect to the quantum field configurations, can be determined within an uncertainty of 30% for C_1 and 50% for $C_{\gamma_3\gamma_5}$ within this setup.

Clearly, this is not satisfactory. But, as we pointed out above, the use of stochastic estimator techniques is still in its infancy. The SESAM result sets the stage of what has to be invested to achieve the goal of calculating disconnected contributions within a few percent uncertainty.

Besides the expected increase of computational power over the next years, improvements of the stochastic estimator technique will help to increase the statistical significance in the calculation of disconnected amplitudes. Promising suggestions along this line can be found in [24] and [25].

References

1. N. Metroprolis, A.W. Rosenbluth, M.N. Rosenbluth, A.H. Teller, and E. Teller: Equation of State Calculations by Fast Computing Machines, J. Chem. Phys.21 (1953)1087; N. Cabibbo and E. Marinari: A New Method for Upadating $SU(N)$ Matrices in Computer Simulations of Gauge Theories, Phys. Lett. 119B (1982)387; K. Fabricius and O. Haan: Heat Bath Method for

the Twisted Eguchi-Kawai Model, Phys. Lett. B143 (1984)459; A. Kennedy and B. Pendleton: Improved Heat Bath Method for Monte Carlo Calculations in Lattice Gauge Theories, Phys. Lett. B156 (1985)393; M. Creutz: Overrelaxation and Monte Carlo Simulation, Phys. Rev. D36 (1987)515.

2. S. Duane, A.D. Kennedy, B.J. Pendleton, and D.Roweth: Hybrid Monte Carlo, Phys. Lett. B 195 (1987)216; M. Lüscher: A New Approach to the Problem of Dynamical Quarks in Numerical Simulations of Lattice QCD, Nucl. Phys. B418 (1994) 637.

3. A.D. Kennedy: The Hybrid Monte Carlo Algorithm on Parallel Computers, Parallel Computing 25 (1999) 1311.

4. For recent reviews see:
G.M. Shore: The 'Proton Spin' Effect: Theoretical Status '97, Nucl. Phys. Proc. Suppl. 64 (1998)167;
H.-J. Cheng: Status of the Proton Spin Problem, Int. J. Mod. Phys. A11 (1996)5109;
M. Anselmino, A. Efremov, and E. Leader: The Theory and Phenomenology of Polarized Deep Inelastic Scattering, Phys. Rep. 261 (1995)1, erratum ibid. 281 399 (1997).

5. G. t'Hooft: Computation of the Quantum Effects due to a Four-Dimensional Pseudoparticle, Phys. Rev. D14 (1976)3432, erratum ibid. D18 2199 (1978); E. Witten: Current Algebra Theorems for the $U(1)$ 'Goldstone Boson', Nucl. Phys. B156 (1979)269; G. Veneziano: U(1) without Instantons, Nucl. Phys. B159 (1979)213.

6. R. Burkhalter: Recent Results from the CP-PACS Collaboration, hep-lat/9810043, Nucl. Phys. B (Proc. Suppl.) 1999, in print.

7. R. Gupta: General Physics Motivations for Numerical Simulations of Quantum Field Theory, Parallel Computing 25 (1999) 1199.

8. J. Gasser, H. Leutwyler, and M.E. Sainio: Sigma Term Update, Phys. Lett. B253(1991)252; J. Gasser, H. Leutwyler, and M.E. Sainio: Form-Factor of the Sigma Term, Phys. Lett. B253 (1991)260.

9. T. Lippert: Parallel SSOR Preconditioning for Lattice QCD, Parallel Computing 25 (1999) 1357.

10. K. Bitar, A.D. Kennedy, R. Horsley, S. Meyer, and P. Rossi: Hybrid Monte Carlo and Quantum Chromodynamics, Nucl. Phys. B313 (1989)348.

11. S.J. Dong and K.F. Liu: Quark Loop Calculations, Nucl. Phys. B(Proc. Suppl.)26 (1992)353.

12. S.J. Dong and K.F. Liu: Stochastic Estimation with $Z(2)$ Noise, Phys. Lett. B328 (1994)130.

13. SESAM Collaboration, S. Güsken, P. Ueberholz, J. Viehoff, N. Eicker, P. Lacock, T. Lippert, K. Schilling, A. Spitz, and T. Struckmann: The Pion Nucleon Sigma Term with Dynamical Wilson Fermions, Phys. Rev. D59, 054504 (1999).

14. SESAM Collaboration, N. Eicker, U. Glässner, S. Güsken, H. Hoeber, T. Lippert, G. Ritzenhöfer, K. Schilling, G. Siegert, A. Spitz, P. Ueberholz, and J. Viehoff: Evaluating Sea Quark Contributions to Flavour-Singlet Operators in Lattice QCD, Phys. Lett. B389 (1996)720.

15. For an overview see Web-page http://www.ncsa.uiuc.edu/Apps/CMP/RNG/

16. M. Fukugita, Y. Kuramashi, M. Okawa, and A. Ukawa: Pion - Nucleon Sigma Term in Lattice QCD, Phys. Rev. D51 (1995)5319.

17. M. Fukugita, Y. Kuramashi, M. Okawa, and A. Ukawa: Proton Spin Structure from Lattice QCD, Phys. Rev. Lett. 75 (1995)2092.

18. S.J. Dong, J.F. Lagaë, and K.F. Liu: Pi N Sigma Term, Anti-S S in Nucleon, and Scalar Form-Factor: A Lattice Study, Phys. Rev. D54 (1996)5496; S.J. Dong and K.F. Liu, Nucl. Phys. B (Proc. Suppl.)42 (1995)322.
19. S.J. Dong, J.F. Lagaë, and K.F. Liu: Flavor Singlet g(A) from Lattice QCD, Phys. Rev. Lett. 75 (1995)2096.
20. SESAM Collab., S. Güsken, P. Ueberholz, J. Viehoff, N. Eicker, T. Lippert, K. Schilling, A. Spitz, and T. Struckmann: The Flavor Singlet Axial Coupling of the Proton with Dynamical Wilson Fermions, preprint WUB98-44, HLRZ 1998-85, hep-lat/9901009, Phys. Rev. D. in print.
21. L. Venkataraman and G. Kilcup: The η' Meson with Staggered Fermions, hep-lat/9711006.
22. F. Rapuano: Physics on APE Computers, Parallel Computing 25 (1999) 1217.
23. W. Wilcox, "Noise Methods for Flavor Singlet Quantities", this volume.
24. Ph. de Forcrand: Monte Carlo Quasi-Heatbath by Approximate Inversion, Phys. Rev. E 59 (1999)3698.
25. SESAM Collaboration, J. Viehoff et al.: News on Disconnected Diagrams, hep-lat/9809130, to be published in Nucl. Phys. (Proc. Suppl.) 1999.

Noise Methods for Flavor Singlet Quantities

Walter Wilcox

Physics Department
Baylor University
Waco TX, USA 76798

Abstract. A discussion of methods for reducing the noise variance of flavor singlet quantities ("disconnected diagrams") in lattice QCD is given. After an introduction, the possible advantage of partitioning the Wilson fermion matrix into disjoint spaces is discussed and a numerical comparison of the variance for three possible partitioning schemes is carried out. The measurement efficiency of lattice operators is examined and shown to be strongly influenced by the Dirac and color partitioning choices. Next, the numerical effects of an automated subtraction algorithm on the noise variance of various disconnected loop matrix elements are examined. It is found that there is a dramatic reduction in the variance of the Wilson point-split electromagnetic currents and that this reduction persists at small quark mass.

1 Introduction

1.1 Motivations

The calculation of flavor singlet quantities, also referred to as disconnected diagrams because the fermion lines are disjoint, is one of the greatest technical challenges left in lattice QCD. Disconnected contributions are present in a wide variety of quantities in strong interaction physics including all baryon form factors, axial operators involving quark spin (Ellis-Jaffe sum rule), hadronic coupling constants and polarizabilities, the p-N sigma term, and propagation functions for various flavor singlet mesons. Such quantities are also present in deep inelastic structure functions measured on the lattice using the operator product expansion, but are not included because of their difficulty and large Monte Carlo error bars. These types of diagrams are difficult to evaluate because exact extractions require many matrix inversions to measure all the background fermion degrees of freedom (including spacetime). Disconnected quark contributions are instead isolated stochastically by a process of applying "noises" to the fermion matrix to project out the desired operator contribution. For an overview of selected aspects of flavor singlet calculations in lattice QCD, see Ref.[1].

1.2 Mathematical Background

Noise methods are based upon projection of the signal using random noise vectors as input. That is, given

$$Mx = \eta,$$

where M is the quark matrix, x is the solution vector and η is the noise vector, with

$$< \eta_i >= 0, < \eta_i \eta_j >= \delta_{ij},$$

where one is averaging over the noise vectors, any inverse matrix element, M_{ij}^{-1}, can then be obtained from

$$< \eta_j x_i >= \sum_k M_{ik}^{-1} < \eta_j \eta_k >= M_{ij}^{-1}.$$

We shall consider two techniques for reducing the noise variance in lattice QCD simulations: partitioning[2] and subtraction methods[3]. Partitioning the noise appropriately, which means most generally a zeroing out of some pattern of noise vector elements, but which will specifically be implemented here by using separate noise source vectors in Dirac and color spaces, can lead to significant reductions in the variance. Subtraction methods, which involve forming new matrix operators which have a smaller variance but the same expectation value (i.e. are "unbiased"), can also be of great help. The key is using a perturbative expansion of the quark matrix as the subtraction matrices which, however, are not unbiased in general and require a separate calculation, either analytical or numerical, to remove the bias. We will see that the various lattice operators have dramatically different behaviors under identical partitioning or subtraction treatments. Both of these methods, partitioning and subtraction, will be treated numerically but the hope is that the numerical results will eventually be "explained" by some simple rules based on the structure of the Wilson matrix.

2 Noise Theory

2.1 Variance Evaluations

Let us review the basics of matrix inversion using noise theory. The theoretical expressions for the expectation value and variance (V) of matrices with various noises are given in Ref.[4]. One has that

$$X_{mn} \equiv \frac{1}{L} \sum_{l=1}^{L} \eta_{ml} \eta_{nl}^*. \tag{1}$$

$(m, n = 1, \ldots, N; l = 1, \ldots, L.)$ We have

$$X_{mn} = X_{nm}^*, \tag{2}$$

and the expectation value,

$$< X_{mn} >= \delta_{mn}. \tag{3}$$

By definition the variance is given by

$$V[Tr\{QX\}] \equiv < |\sum_{m,n} q_{mn} X_{nm} - Tr\{Q\}|^2 > . \tag{4}$$

The variance may be evaluated as,

$$V[Tr\{QX\}] = \sum_{m \neq n} (< |X_{nm}|^2 > |q_{mn}|^2 \tag{5}$$

$$+ q_{mn} q_{nm}^* < (X_{mn})^2 >) + \sum_n < |X_{nn} - 1|^2 > |q_{nn}|^2.$$

2.2   Real Noises

For a general real noise,

$$< |X_{mn}|^2 >= \frac{1}{L}, \tag{6}$$

$$< (X_{mn})^2 >= \frac{1}{L}, \tag{7}$$

for $m \neq n$ so that

$$V[Tr\{QX_{\text{real}}\}] = \frac{1}{L} \sum_{m \neq n} (|q_{mn}|^2 + q_{mn} q_{nm}^*) \tag{8}$$

$$+ \sum_n < |X_{nn} - 1|^2 > |q_{nn}|^2.$$

The case of real $Z(2)$ has Eqs.(6) and (7) holding for $m \neq n$, but also

$$< |X_{nn} - 1|^2 >= 0. \tag{9}$$

This shows that

$$V[Tr\{QX_{Z(2)}\}] \leq V[Tr\{QX_{\text{real}}\}]. \tag{10}$$

Thus, $Z(2)$ noise has the lowest variance of any real noise.

2.3   General Z(N) Noise

For general $Z(N)$ ($N \geq 3$) noise we have a different situation. One has that

$$< |X_{mn}|^2 >= \frac{1}{L},$$ (11)

$$< (X_{mn})^2 >= 0,$$ (12)

for $m \neq n$, and again

$$< |X_{nn} - 1|^2 >= 0.$$ (13)

Thus

$$V[Tr\{QX_{Z(N)}\}] = \frac{1}{L} \sum_{m \neq n} |q_{mn}|^2,$$ (14)

and the variance relationship of $Z(2)$ and $Z(N)$ is not fixed for a general matrix Q. The reason for the difference in Eqs.(7) and (12) is that the square of an equally weighted distribution of $Z(2)$ elements is not itself uniformly distributed (always 1), whereas the square of a uniformly weighted $Z(N)$ distribution for $N \geq 3$ is also uniformly distributed. However, if the phases of q_{mn} and q_{nm}^* are uncorrelated, then $V[Tr\{QX_{Z(2)}\}] \approx V[Tr\{QX_{Z(N)}\}]$, ($N \geq 3$) which, we will see, is apparently the case for the operators studied here.

3   Partitioning the Problem

3.1   Basic Idea

By "partitioning" I mean replacing the single noise vector problem,

$$Mx = \eta,$$ (15)

which yields a complete output column, $\sum_k M_{ik}^{-1}\eta_k$, from a single input noise vector, η, with a problem

$$Mx^p = \eta^p, p = 1, \ldots P,$$ (16)

where the η^p have many zeros corresponding to some partitioning scheme. In this latter case it takes P inverses to produce a complete measurement or sampling of a column of M^{-1}.

For the unpartitioned problem (for $Z(N), N \geq 3$, say)

$$V[Tr\{QX\}] = \frac{1}{L} \sum_{m \neq n} |q_{mn}|^2,$$ (17)

$$\equiv \frac{1}{L} N(N - 1) < |q|^2 >,$$

where I have defined the average absolute squared off diagonal matrix element, $< |q|^2 >$. For the partitioned problem the total variance includes a sum on p,

$$\sum_{p=1}^{P} V[Tr\{QX_p\}] = \frac{1}{L} \sum_{p=1}^{P} \sum_{m_p \neq n_p} |q_{m_p n_p}|^2, \tag{18}$$

$$\equiv \frac{1}{L} N(\frac{N}{P} - 1) < |q_P|^2 > .$$

In order for this method to pay off in terms of computer time, one needs that

$$\sum_{p=1}^{P} V[Tr\{QX_p\}] \leq \frac{1}{P} V[Tr\{QX\}], \tag{19}$$

$$\Rightarrow < |q_P|^2 > \leq \left(\frac{N-1}{N-P} \right) < |q|^2 > . \tag{20}$$

The goal of partitioning is to avoid some of the large off-diagonal matrix elements so that in spite of doing P times as many inverses, a smaller variance is produced for the same amount of computer time. The spaces partitioned can be space-time, color or Dirac or some combination. I have found that partitioning in Dirac and color spaces can strongly affect the results.

3.2   Simulation Description

I consider all local operators, $\bar{\psi}(x)\Gamma\psi(x)$, as well as point-split versions of the vector and axial vector operators. This means 16 local operators and 8 point-split ones, making a total of 24, which are listed below. For each operator there are both real and imaginary parts, but in each case one may show via the quark propagator identity $S = \gamma_5 S^\dagger \gamma_5$, that only the real *or* the imaginary part of each local or nonlocal operator is nonzero on a given configuration for each space-time point. However, this identity is not respected exactly by noise methods, so the cancellations are actually only approximate configuration by configuration. However, the knowledge that one part is purely noise allows one to simply drop that part in the calculations, thus reducing the variance without biasing the answer.

The operators I consider are:

- Scalar: $Re[\bar{\psi}(x)\psi(x)]$
- Local Vector: $Im[\bar{\psi}(x)\gamma_\mu\psi(x)]$
- Point-Split Vector:
   $\kappa Im[\bar{\psi}(x + a_\mu)(1 + \gamma_\mu)U_\mu^\dagger(x)\psi(x) - \bar{\psi}(x)(1 - \gamma_\mu)U_\mu(x)\psi(x + a_\mu)]$
- Pseudoscalar: $Re[\bar{\psi}(x)\gamma_5\psi(x)]$
- Local Axial: $Re[\bar{\psi}(x)\gamma_5\gamma_\mu\psi(x)]$
- Point-Split Axial:

$$\kappa Re[\bar{\psi}(x + a_\mu)\gamma_5\gamma_\mu U_\mu^\dagger(x)\psi(x) + \bar{\psi}(x)\gamma_5\gamma_\mu U_\mu(x)\psi(x + a_\mu)]$$

- Tensor: $Im[\bar{\psi}(x)\sigma_{\mu\nu}\psi(x)]$

I actually consider the zero momentum version of these operators, summed over both space and time.

The sample noise variance in M quantities x_i is given by the standard expression:

$$V_{noise} = \frac{1}{M-1}\sum_{i=1,M}(x_i - \bar{x})^2 \qquad (21)$$

What I concentrate on here are the relative variances between the different methods. Since the squared noise error in a single configuration is given by

$$\sigma^2_{noise} = \frac{V_{noise}}{M}, \qquad (22)$$

the ratio of variances gives a direct measure of the multiplicative ratio of noises, and thus the computer time, necessary to achieve the same noise error. However, the variance itself does not take into account the extra P inverses done when the problem is partitioned. In order to measure the relative efficiency of different partitionings, I form what I call pseudo-efficiencies ratios ("PE"), which are defined by

$$\text{PE}(\frac{\text{method1}}{\text{method2}}) \equiv \frac{P_{method1}(V_{noise})^{method1}}{P_{method1}(V_{noise})^{method2}}, \qquad (23)$$

where P_{method} are the number of partitions required by the method. I refer to these ratios as "pseudo" efficiencies since I do a fixed number of iterations for all of the operators I consider. It could very well be that different methods will require significantly different numbers of iterations of conjugate-gradient or minimum residual for the same level of accuracy.

One desires to find the lowest PE ratio for a given operator. [1]

I display PE ratio results for Wilson fermions in a $16^3 \times 24$, $\beta = 6.0$ lattice with $\kappa = 0.148$ in Table 1, which follows on the next page. (Part of this Table also appeared in Ref.[2].) I will examine 3 partitionings:

- $Z(2)$ unpartitioned ("$P = 1$ $Z(2)$");
- $Z(2)$ Dirac partitioned ("$P = 4$ $Z(2)$");
- $Z(2)$ Dirac and color partitioned ("$P = 12$ $Z(2)$").

---

[1] Of course an evaluation via N partitioning on an $N \times N$ matrix yields a PE numerator factor of zero relative to other methods since the variance is exactly zero in this case. This is of course prohibitively expensive, but it suggests that more efficient partitionings are possible for large computer budgets. Thanks to M. Peardon for bringing this point out.

My Dirac gamma matrix representation is:

$$\gamma_i = \begin{pmatrix} 0 & \sigma_i \\ \sigma_i & 0 \end{pmatrix}, \gamma_4 = \begin{pmatrix} 1 & 0 \\ 0 & -1 \end{pmatrix}, \gamma_5 = \gamma_1 \gamma_2 \gamma_3 \gamma_4 = \begin{pmatrix} 0 & -i \\ i & 0 \end{pmatrix}. \tag{24}$$

In Table 1 I list the relative PEs of the two partitioned methods relative to the unpartitioned case. Referring to the above list of the real or imaginary parts of operators my notation here is as follows: "Scalar" stands for the operator $\bar{\psi}\psi$, "Local Vector 1" for example stands for the operator $\bar{\psi}\gamma_1\psi$, "P-S Vector 1" stands for the 1 component of the point split vector current, "pseudoscalar" stands for $\bar{\psi}\gamma_5\psi$, "Local Axial 1" stands for the operator $\bar{\psi}\gamma_5\gamma_1\psi$, "P-S Axial 1" stands for the point split axial 1 component, and "Tensor 41" stands for example for the operator $\bar{\psi}\sigma_{41}\psi$.

There are extremely large variations in the behaviors of the operators listed in Table 1 under identical partitionings. Of the partitionings considered it is most efficient to calculate scalar and vector operators with an unpartitioned simulation. On the other hand, it is far more efficient to calculate the pseudoscalar in a Dirac and color partitioned manner. Notice the entries for the 1,2 components of the axial current (both local and point split) do not behave like the 3,4 components under pure Dirac partitioning, but they do when Dirac and color partitionings are combined. Four of the tensor operators respond best to a pure Dirac partitioning, while the other two prefer a partitioning in Dirac and color spaces combined. The ratio of the largest to the smallest entry in the right hand column is about 800!

As pointed out in Section 2, the variance of $Z(2)$ and $Z(N)$ ($N \geq 3$) noises are in general different. For this reason I also investigated partitioning using $Z(4)$ as well as volume (gauge variant) noises, but there do not seem to be large factors to be gained relative to the $Z(2)$ case. The SESAM collaboration also has seen the efficacy of partitioning (in Dirac space) for axial operators[5].

4   Perturbative Noise Subtraction

4.1   Description of Algorithm

Consider $\tilde{Q}$ such that

$$< Tr\{\tilde{Q}X\} >= 0. \tag{25}$$

One can then form

$$< Tr\{(Q - \tilde{Q})X\} >=< Tr\{QX\} > . \tag{26}$$

However,

$$V[Tr\{(Q - \tilde{Q})X\}] \neq V[Tr\{QX\}]. \tag{27}$$

Table 1. The pseudoefficiency (PE) ratios associated with the methods indicated.

Operator	PE($\frac{P=4\ Z(2)}{P=1\ Z(2)}$)	PE($\frac{P=12\ Z(2)}{P=1\ Z(2)}$)
Scalar	2.83 ± 0.47	10.9 ± 2.5
Local Vector 1	2.38 ± 0.65	8.71 ± 1.8
Local Vector 2	2.50 ± 0.53	12.1 ± 2.9
Local Vector 3	3.60 ± 1.00	11.4 ± 2.4
Local Vector 4	3.41 ± 0.60	16.3 ± 3.4
P-S Vector 1	2.63 ± 0.56	9.94 ± 2.2
P-S Vector 2	2.27 ± 0.44	11.0 ± 2.3
P-S Vector 3	3.52 ± 0.74	11.4 ± 1.5
P-S Vector 4	3.87 ± 0.49	15.5 ± 4.2
Pseudoscalar	0.698 ± 0.15	0.0201 ± 0.0043
Local Axial 1	0.114 ± 0.021	0.144 ± 0.029
Local Axial 2	0.126 ± 0.020	0.146 ± 0.037
Local Axial 3	1.13 ± 0.19	0.162 ± 0.038
Local Axial 4	2.26 ± 0.24	0.187 ± 0.035
P-S Axial 1	0.167 ± 0.040	0.151 ± 0.018
P-S Axial 2	0.110 ± 0.032	0.114 ± 0.028
P-S Axial 3	1.67 ± 0.35	0.186 ± 0.049
P-S Axial 4	1.85 ± 0.21	0.238 ± 0.036
Tensor 41	1.07 ± 0.30	0.295 ± 0.049
Tensor 42	0.345 ± 0.076	0.0889 ± 0.011
Tensor 43	1.32 ± 0.43	0.398 ± 0.13
Tensor 12	1.12 ± 0.25	0.376 ± 0.066
Tensor 13	0.116 ± 0.024	0.363 ± 0.053
Tensor 23	0.0314 ± 0.0058	0.0751 ± 0.016

As we have seen for $Z(N)$ $(N \geq 2)$, the variance comes exclusively from off diagonal entries. So, the trick is to try to find matrices $\tilde{Q}$ which are traceless (so they do not affect the expectation value) but which mimic the off-diagonal part of Q as much as possible to reduce the variance.

The natural choice is simply to choose as $\tilde{Q}$ the *perturbative* expansion of the quark matrix. Formally, one has $(I, J = \{x, a, \alpha\})$

$$(M^{-1})_{IJ} = \frac{1}{\delta_{IJ} - \kappa P_{IJ}}, \tag{28}$$

where

$$P_{IJ} = \sum_{\mu} [(1 + \gamma_\mu) U_\mu(x) \delta_{x,y-a_\mu} + (1 - \gamma_\mu) U_\mu^\dagger(x - a_\mu) \delta_{x,y+a_\mu}]. \tag{29}$$

Expanding this in κ gives the perturbative (or hopping parameter) expansion,

$$M_p^{-1} = I + \kappa P + \kappa^2 P^2 + \kappa^3 P^3 + \cdots . \tag{30}$$

One constructs $< \eta_j (M_p^{-1})_{ik} \eta_k >$ and subtracts it from $< \eta_j M_{ik}^{-1} \eta_k >$, where η is the noise vector. Constructing $< \eta_j (M_p^{-1})_{ik} \eta_k >$ is an iterative process and is easy to code and extend to higher powers on the computer. I will iterate up to 10th order in κ.

One can insert coefficients in front of the various terms and vary them to find the minimum in the variance, but such coefficients are seen to take on their perturbative value, at least for high order expansions[6]. However, see also Ref.[7] where in low orders there is apparently an advantage to this procedure. Interestingly, significant subtraction improvements occur in some operators even in 0th order (point split vectors and two tensor operators.)

For a given operator, $\mathcal{O}$, the matrix $\mathcal{O}M_p^{-1}$ encountered in the context of $< \bar{\psi}\mathcal{O}\psi >= -Tr(\mathcal{O}M^{-1})$ is not traceless. In other words, one must re-add the perturbative trace, subtracted earlier, to get the full, unbiased answer. How does one calculate the perturbative part? The exact way is of course to explicitly construct all the gauge invariant paths (up to a given κ order) for a given operator. Another approach is to subject the perturbative contribution to a separate Monte Carlo estimation. This is the approach taken here. Local operators require perturbative corrections starting at 4th order (except a trivial correction for $\bar{\psi}\psi$ at zeroth order) and point split ones require corrections starting instead at 3rd order. Because one is removing the bias (perturbative trace) by a statistical method, I refer to this as a "statistically unbiased" method. Some efficiency considerations in carrying out this procedure will be discussed in Section 5. Other versions of subtraction methods in the context of lattice evaluations of disconnected diagrams may be found in Refs.[8] and [9].

4.2 Numerical Results

I am carrying out this numerical investigation in an unpartitioned sense
($P = 1$). The operators which respond best to this partitioning, as discussed
previously, are the scalar and local and point-split vector currents, and attention will be limited here to these cases. The effect of combining the partition
and subtraction methods has not yet been investigated. I show the ratio of
unsubtracted variance divided by subtracted variance, $\mathbf{V}_{unsub}/\mathbf{V}_{sub}$ in Figs.
1 and 2. Factors larger than one give the multiplicative gain in computer time
one is achieving. The lattices are again Wilson $16^3 \times 24$, $\beta = 6.0$.

Notice the approximate linear rise in the variance ratio as a function
of subtraction order for the point-split vector charge density at both $\kappa =
0.148$ and 0.152, Figs. 1 and 2 respectively. Also notice that even at $S = 0$
(subtracting the Kronecker delta) there is a reduction in the variance. The
slope of the subtraction graph at $\kappa = 0.148$ is about 3.5; the slope at $\kappa = 0.152$
is reduced to a little under 3.0. Although I do not show the results here, the
same linear behavior is evident in $\bar{\psi}\psi$ and the local vector operators although
their slopes are considerably smaller.

My final results are summarized in Fig. 3, which gives the reduction in
the variance in the scalar and vector operators after a 10th order subtraction
has been made at $\kappa = 0.148$. It is not known why the point split vector
current responds the best to subtraction. The 10th order point-split vector,
local vector, and scalar variance ratios change from ~ 35, ~ 12, and ~ 10 at
$\kappa = 0.148$, to $\sim 25 \sim 10$, and ~ 5 at $\kappa = 0.152$, respectively. These are all
zero momentum operators. Although the results are not shown, I have found
essentially identical results to the above for momentum transformed data,
necessary for disconnected form factors. Perturbative subtraction methods
will thus be extremely useful in lattice evaluations of nucleon strangeness
form factors using the point split (conserved) form of the vector current.

5 Efficiency Considerations

5.1 Fixed Noise Case

Let me close with some simple observations regarding statistical errors in
flavor singlet Monte Carlo simulations. There are two sources of variance in
such simulations: gauge configuration and noise. Given N configurations and
M noises per configuration, the final error bar on a given operator is given
by

$$\sigma = \sqrt{\frac{V_{noise}}{NM} + \frac{V_{gauge}}{N}}, \tag{31}$$

where V_{guage} and V_{noise} are the gauge configuration and noise variances. For
fixed NM (total number of noises), it is clear that Eq.(22) is minimized by
taking $M = 1$. Thus, in this situation it is best to use a single noise per

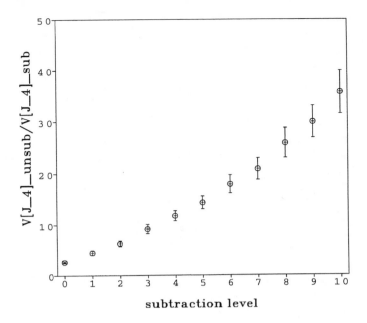

Fig. 1. Effect of the level of perturbative subtraction, up to tenth order in κ, on the ratio of unsubtracted divided by subtracted noise variance for the zero momentum point-split (conserved) charge density operator, J_4, at $\kappa = 0.148$.

configuration. This simple result is modified by real world considerations of overheads. For example, if one assumes that there is an overhead associated with generating configurations and fixes instead the total amount of computer time,

$$T = NM + G_N N, \tag{32}$$

where G_N is the appropriately scaled configuration generation time overhead, then one finds instead that

$$M = \frac{S_{noise}}{S_{gauge}} \sqrt{G_N}, \tag{33}$$

is the best choice. Note that the ratio S_{noise}/S_{gauge} can have a wide range of values for various operators, and one is no longer guaranteed that $M = 1$ is optimal.

5.2   Fixed Configuration Case

Another common real world situation is where N, the number of configurations, is fixed. In the context of the perturbative subtraction algorithm, one

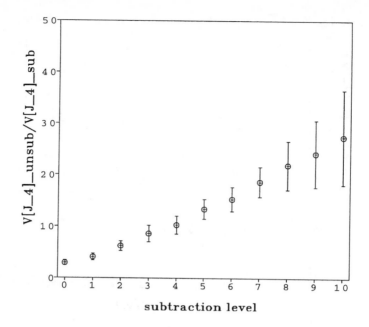

Fig. 2. Same as Fig. 1 but for $\kappa = 0.152$.

should now maximize the number of effective noises for a given computer budget. The effective number of noises is

$$M_{eff} = M(X + S\Delta s), \tag{34}$$

where M_{eff} replaces M in Eq.(22). (M retains its meaning as the actual number of gauge field noises.) S is the subtraction order, ranging from 0 to 10 in Figs. 1 and 2, and Δs is the slope. X is the factor one obtains from this method at $S = 0$, without extra overhead. (One sees in Figs. 1 and 2 that this factor is about 2 for the point split charge operator.) I am assuming that the reduction in the variance is approximately linear in subtraction order S. S is imagined to be a continuously variable quantity. The total time per gauge field is

$$T_N = (MT_M + ST_S), \tag{35}$$

which is kept fixed as M_{eff} is varied. T_M is the noise time overhead and T_S is the subtraction time overhead. The optimum choices for S and M are now

$$S = \frac{T_N}{2T_S} - \frac{X}{2\Delta s}, \tag{36}$$

$$M = \frac{T_N}{2T_M} + \frac{XT_S}{2\Delta sT_M}, \tag{37}$$

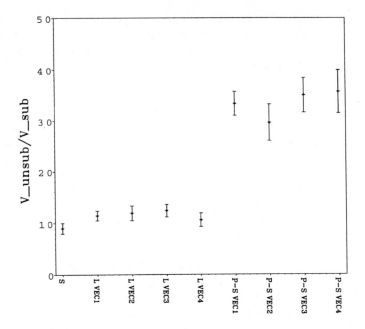

Fig. 3. Graphical presentation of the unsubtracted variance divided by subtracted variance of 9 different lattice operators after tenth order subtraction at $\kappa = 0.148$ for Wilson fermions. (Abbreviations used for operators: S=scalar; L VEC = local vector; P-S VEC = point split vector).

resulting in

$$M_{eff} = \frac{\Delta s T_N^2}{4 T_S T_M} + \frac{T_N X}{2 T_M} - \frac{X^2 T_S}{4 \Delta s T_M}. \tag{38}$$

The interesting aspect of this last result is that the effective number of noises, M_{eff}, is now quadratic in T_N. This is a consequence of our observation that the slopes in Figs. 1 and 2 are approximately linear in S. The immediate implication is that for large T_N the noise error bar can be made to vanish like the inverse of the simulation time rather than as the usual inverse square root, at least in the range of the existing linear behavior.

These equations are also helpful when one has an exact analytical representation of the trace of the perturbative series up to some order, S_{exact}, making T_S zero. Then, by comparing the M_{eff} values in the two cases, one may find the lowest value of S, S_{lowest}, such that $M_{eff}^{lowest} \geq M_{eff}^{exact}$. For example, when $X \approx 0$ and for common values of T_N and T_M in the two simulations, one has

$$S_{lowest} = 2 S_{exact}. \tag{39}$$

That is, for the extra Monte Carlo overhead to pay off, one must attempt to subtract to at least twice as high an order in κ as the exact evaluation. Since subtraction is always exact for nonlocal operators up to second order and for local ones up to third order, making $S_{lowest} \geq 6$ will usually result in a more efficient simulation than the default exact one.

6 Summary

We have seen that significant savings in computer resources may be obtained by partitioning the Wilson matrix appropriately. An efficient partitioning reduces the variance of an operator by leaving out the largest off-diagonal matrix elements of the quark propagator so that in spite of having to do more inversions, a smaller variance is produced for the same amount of computer time. A numerical investigation in Dirac and color spaces revealed efficient partitionings for 24 local and nonlocal operators summarized near the end of Section 3.

We have also seen that large time savings are possible using subtraction methods for selected operators in the context of unpartitioned noise simulations. This method was shown to be effective for the scalar and local vector currents, but most effective for the point-split vector currents. Since perturbative subtraction is based on the hopping parameter expansion of the quark propagator, such methods can become less effective at lower quark masses, although we found the variance reduction was still quite significant for the point-split vector operator at $\kappa = 0.152$. Similar methods can be devised for other operators (axial, pseudoscalar, tensor) by implementing these ideas in the context of Dirac/color partitioned noise methods.

There are still a number of open questions here. The reasons for the strange partitioning patterns found in Section 3 are not known. In addition, the reason why the variance of some operators respond more sensitively to perturbative subtraction than others is obscure. These questions are important because their answers could lead one to better simulation methods. Another question is how far the linear slopes in Figs. 1 and 2 persist at high subtraction orders. Since the perturbative expansion of the Wilson matrix does not converge at small quark mass, the slope of such curves probably levels off at high enough S. We have seen, however, that before this leveling off occurs the number of effective noises grows quadratically in the simulation time. It was pointed out that this implies that the noise error bar can be made to vanish like the inverse of the simulation time in the range of the existing linear behavior.

7 Acknowledgements

This work was supported in part by National Science Foundation Grant No. 9722073, and generous grants of computer time on the SGI Origin 200 computer from the National Center for Supercomputing Applications (NCSA) in Urbana/Champaign, Illinois. I thank the University of Kentucky Department of Physics and Astronomy and the Special Research Centre for the Subatomic Structure of Matter, University of Adelaide for their hospitality and invitations while portions of this work were underway. I also thank the conference organizers at the University of Wuppertal for their invitation to this workshop.

References

1. Güsken, S.: Flavor singlet phenomena in lattice QCD. hep-lat/9906034
2. Wilcox, W. and Lindsay, B.: Disconnected loop noise methods in lattice QCD. Nucl. Phys. **B** (Proc. Suppl.) **63A-C** (1998) 973–975
3. Wilcox, W.: Perturbative subtraction methods. Nucl. Phys. **B** (Proc. Suppl., to appear) hep-lat/9908001
4. Bernardson, S., McCarty, P., and Thron, C.: Monte carlo methods for estimating linear combinations of inverse matrix entries in lattice QCD. Comp. Phys. Comm. **78** (1994) 256–264
5. SESAM Collaboration, J. Viehoff et. al.: Improving stochastic estimator techniques for disconnected diagrams. Nucl. Phys. **B** (Proc. Suppl.) **63** (1998) 269–271
6. Thron, C., Dong, S-J, Liu, K-F, Ying, H-P: Padé-Z2 estimator of determinants. Phys. Rev. **D57** (1998) 1642–1653
7. Mathur, N., Dong, S-J, Liu, K-F and Mukhopadhyay, N.: Proton spin content from lattice QCD. Nucl. Phys. **B** (Proc. Suppl., to appear) hep-lat/9908052
8. Michael C., Foster, M., and McNeile, C.: Flavor singlet pseudoscalar and scalar mesons. Nucl. Phys. **B** (Proc. Suppl., to appear) hep-lat/9909036
9. SESAM collaboration, J. Viehoff et. al.: News on disconnected diagrams. Nucl. Phys. **B** (Proc. Suppl.) **73** (1999) 856–858

A Noisy Monte Carlo Algorithm with Fermion Determinant

Keh-Fei Liu

Dept. of Physics and Astronomy
Univ. of Kentucky
Lexington, KY 40506, USA

Abstract. We propose a Monte Carlo algorithm which accommodates an unbiased stochastic estimates of the fermion determinant and is exact. This is achieved by adopting the Metropolis accept/reject steps for the update of both the dynamical and noise configurations. We demonstrate how this algorithm works even for large noises. We also discuss the Padé - Z_2 estimates of the fermion determinant as a practical way of estimating the Trln of the fermion matrix.

1 A Noisy Monte Carlo Algorithm

There are problems in physics which involve extensive quantities such as the fermion determinant which require V^3 steps to compute exactly. Problems of this kind with large volumes are not numerically applicable with the usual Monte Carlo algorithm which require the exact evaluation of the probability ratios in the accept/reject step. To address this problem, Kennedy and Kuti [1] proposed a Monte Carlo algorithm which admits stochastically estimated transition probabilities as long as they are unbiased. But there is a drawback with this algorithm. The probability could lie outside the interval between 0 and 1 since it is estimated stochastically. This probability bound violation will destroy detailed balance and lead to systematic bias. To control the probability violation with a large noise ensemble can be costly.

We propose a noisy Monte Carlo algorithm which avoids this difficulty with two Metropolis accept/reject steps. Let us consider a model with Hamiltonian $H(U)$ where U collectively denotes the dynamical variables of the system. The major ingredient of the new approach is to transform the noise for the stochastic estimator into stochastic variables. The partition function of the model can be written as

$$Z = \int [DU]\, e^{-H(U)}$$

$$= \int [DU][D\xi] P_\xi(\xi)\, f(U, \xi). \tag{1}$$

where $f(U, \xi)$ is an unbiased estimator of $e^{-H(U)}$ from the stochastic variable ξ and P_ξ is the probability distribution for ξ.

The next step is to address the lower probability-bound violation. One first notes that one can write the expectation value of the observable O as

$$\langle O \rangle = \int [DU][D\xi] \, P_\xi(\xi) \, O(U) \, \text{sign}(f) \, |f(U,\xi)|/Z, \tag{2}$$

where $sign(f)$ is the sign of the function f. After redefining the partition function to be

$$Z = \int [DU][D\xi] P_\xi(\xi) \, |f(U,\xi)|, \tag{3}$$

which is semi-positive definite, the expectation of O in Eq. (2) can be rewritten as

$$\langle O \rangle = \langle O(U) \, \text{sign}(f) \rangle / \langle \text{sign}(f) \rangle. \tag{4}$$

As we see, the sign of $f(U,\xi)$ is not a part of the probability any more but a part in the observable. Notice that this reinterpretation is possible because the sign of $f(U,\xi)$ is a state function which depends on the configuration of U and ξ.

It is clear then, to avoid the problem of lower probability-bound violation, the accept/reject criterion has to be factorizable into a ratio of the new and old probabilities so that the sign of the estimated $f(U,\xi)$ can be absorbed into the observable. This leads us to the Metropolis accept/reject criterion which incidentally cures the problem of upper probability-bound violation at the same time. It turns out two accept/reject steps are needed in general. The first one is to propose updating of U via some procedure while keeping the stochastic variables ξ fixed. The acceptance probability P_a is

$$P_a(U_1, \xi \to U_2, \xi) = \min\left(1, \frac{|f(U_2, \xi)|}{|f(U_1, \xi)|}\right). \tag{5}$$

The second accept/reject step involves the refreshing of the stochastic variables ξ according to the probability distribution $P_\xi(\xi)$ while keeping U fixed. The acceptance probability is

$$P_a(U, \xi_1 \to U, \xi_2) = \min\left(1, \frac{|f(U, \xi_2)|}{|f(U, \xi_1)|}\right). \tag{6}$$

It is obvious that there is neither lower nor upper probability-bound violation in either of these two Metropolis accept/reject steps. Furthermore, it involves the ratios of separate state functions so that the sign of the stochastically estimated probability $f(U,\xi)$ can be absorbed into the observable as in Eq. (4).

Detailed balance can be proven to be satisfied and it is unbiased [2]. Therefore, this is an exact algorithm. We have tested this noisy Monte Carlo (NMC) on a 5-state model which is the same used in the linear algorithm [1]

Table 1. Data for the average energy obtained by NMC and the linear algorithm [1]. They are obtained with a sample size of one million configurations each. Var is the variance of the noise estimator for e^{-H} in NMC and $e^{\Delta H}$ in the linear algorithm. Negative Sign denotes the percentage of times when the sign of the estimated probability is negative in NMC. Low/High Vio. denotes the percentage of times when the low/high probability-bound is violated in the linear algorithm. The exact average energy is 0.180086.

Var	NMC	Negative Sign	Linear	Low Vio.	High Vio.
0.001	0.17994(14)	0%	0.18024(14)	0%	0%
0.002	0.18016(14)	0%	0.17994(14)	0%	0%
0.005	0.18017(14)	0%	0.17985(14)	0%	0%
0.008	0.17993(14)	0%	0.17997(14)	0%	0%
0.01	0.18008(14)	0%	0.17991(14)	0%	0%
0.06	0.17992(14)	0.008%	0.17984(14)	0.001%	0.007%
0.1	0.17989(14)	0.1%	0.17964(14)	0.1%	0.3%
0.2	0.18015(15)	1.6%	0.18110(13)	1%	1%
0.5	0.1800(3)	5%	0.1829(1)	3%	4%
1.0	0.1798(4)	12%	0.1860(1)	6%	7%
5.0	0.1795(6)	28%	0.1931(1)	13%	13%
6.5	0.1801(5)	30%	0.1933(1)	13%	14%
10.0	0.1799(9)	38%	-	-	-
15.0	0.1798(9)	38%	-	-	-
20.0	0.1803(11)	39%	-	-	-
30.0	0.1800(13)	41%	-	-	-
50.0	0.1794(17)	44%	-	-	-

for demonstration. Here, $P_c(U_1 \rightarrow U_2) = \frac{1}{5}$ and we use Gaussian noise to mimic the effects of the noise in the linear algorithm and the stochastic variables ξ in NMC. We calculate the average energy with the linear algorithm and the NMC. Some data are presented in Table 1. Each data point is obtained with a sample of one million configurations. The exact value for the average energy is 0.180086.

We first note that as long as the variance of the noise is less than or equal to 0.06, the statistical errors of both the NMC and linear algorithm stay the same and the results are correct within two σ. To the extent that the majority of the numerical effort in a model is spent in the stochastic estimation of the probability, this admits the possibility of a good deal of saving through the reduction of the number of noise configurations, since a poorer estimate of the probability with less accuracy works just as well. As the variance becomes larger than 0.06, the systematic bias of the linear algorithm sets in and eventually becomes intolerable, while there is no systematic bias in the NMC results. In fact, we observe that the NMC result is correct even when the percentage of negative probability reaches as high as 44%, although the statistical fluctuation becomes larger due to the fact that the negative sign appears more frequently. We should remark that the Metropolis acceptance rate is about 92% for the smallest noise variance. It decreases to 85% when the variance is 0.1 and it drops eventually to 78% for the largest variance 50.0. Thus, there is no serious degrading in the acceptance rate when the variance of the noise increases.

We further observe that the variance of the NMC result does not grow as fast as the variance of the noise. For example, the variance of the noise changes by a factor of 833 from 0.06, where the probability-bound violation starts to show up in the linear algorithm, to 50.0. But the variance of the NMC result is only increased by a factor of $(0.0017/0.00014)^2 = 147$. Thus, if one wants to use the linear algorithm to reach the same result as that of NMC and restricts to configurations without probability-bound violations, it would need 833 times the noise configurations to perform the stochastic estimation in order to bring the noise variance from 50.0 down to 0.06 but 147 times less statistics in the Monte Carlo sample. In the case where the majority of the computer time is consumed in the stochastic estimation, it appears that NMC can be more economical than the linear algorithm.

2   Noisy Monte Carlo with Fermion Determinant

One immediate application of NMC is lattice QCD with dynamical fermions. The action is composed of two parts – the pure gauge action $S_g(U)$ and a fermion action $S_F(U) = -\text{Tr}\ln M(U)$. Both are functionals of the gauge link variables U.

To find out the explicit form of $f(U,\xi)$, we note that the fermion determinant can be calculated stochastically as a random walk process [3]

$$e^{\text{Tr}\ln M} = 1 + \text{Tr}\ln M(1 + \frac{\text{Tr}\ln M}{2}(1 + \frac{\text{Tr}\ln M}{3}(...))) . \qquad (7)$$

This can be expressed in the following integral

$$e^{\text{Tr}\ln M} = \int \prod_{i=1}^{\infty} d\eta_i\, P_\eta(\eta_i) \int_0^1 \prod_{n=2}^{\infty} d\rho_n$$
$$[1 + \eta_1^\dagger \ln M\eta_1(1 + \theta(\rho_2 - \frac{1}{2})\eta_2^\dagger \ln M\eta_2(1 + \theta(\rho_3 - \frac{2}{3})\eta_3^\dagger \ln M\eta_3(...], \qquad (8)$$

where $P_\eta(\eta_i)$ is the probability distribution for the stochastic variable η_i. It can be the Gaussian noise or the Z_2 noise ($P_\eta(\eta_i) = \delta(|\eta_i| - 1)$ in this case). The latter is preferred since it has the minimum variance [4]. ρ_n is a stochastic variable with uniform distribution between 0 and 1. This sequence terminates stochastically in finite time and only the seeds from the pseudo-random number generator need to be stored in practice. The function $f(U,\eta,\rho)$ (Note the ξ in Eq. (1) turns into two stochastic variables η and ρ here) is represented by the part of the integrand between the the square brackets in Eq. (8). One can then use the efficient Padé-Z_2 algorithm [5] to calculate the $\eta_i \ln M \eta_i$ in Eq. (8). We shall discuss this in the next section.

Finally, there is a practical concern that $\mathrm{Tr} \ln M$ can be large so that it takes a large statistics to have a reliable estimate of $e^{\mathrm{Tr} \ln M}$ from the series expansion in Eq. (8). In general, for the Taylor expansion $e^x = \sum x^n/n!$, the series will start to converge when $x^n/n! > x^{n+1}/(n+1)!$. This happens at $n = x$. For the case $x = 100$, this implies that one needs to have more than 100! stochastic configurations in the Monte Carlo integration in Eq. (8) in order to have a convergent estimate. Even then, the error bar will be very large. To avoid this difficulty, one can implement the following strategy. First one note that since the Metropolis accept/reject involves the ratio of exponentials, one can subtract a universal number x_0 from the exponent x in the Taylor expansion without affecting the ratio. Second one can use a specific form of the exponential to diminish the value of the exponent. In other words, one can replace e^x with $(e^{(x-x_0)/N})^N$ to satisfy $|x - x_0|/N < 1$. The best choice for x_0 is $\bar{x}$, the mean of x. In this case, the variance of e^x becomes $e^{\delta^2/N} - 1$.

3   The Padé – Z_2 Method of Estimating Determinants

Now we shall discuss a very efficient way of estimating the fermion determinant stochastically [5].

3.1   Padé approximation

The starting point for the method is the Padé approximation of the logarithm function. The Padé approximant to $\log(z)$ of order $[K, K]$ at z_0 is a rational function $N(z)/D(z)$ where deg $N(z) = $ deg $D(z) = K$, whose value and first $2K$ derivatives agree with $\log z$ at the specified point z_0. When the Padé approximant $N(z)/D(z)$ is expressed in partial fractions, we obtain

$$\log z \approx b_0 + \sum_{k=1}^{K} \left(\frac{b_k}{z + c_k} \right), \tag{9}$$

whence it follows

$$\log \det \mathbf{M} = \mathrm{Tr} \, \log \mathbf{M} \approx b_0 \mathrm{Tr} \mathbf{I} + \sum_{k=1}^{K} b_k \cdot \mathrm{Tr}(\mathbf{M} + c_k \mathbf{I})^{-1}. \tag{10}$$

The Padé approximation is not limited to the real axis. As long as the function is in the analytic domain, i. e. away from the cut of the log, say along the negative real axis, the Padé approximation can be made arbitrarily accurate by going to a higher order $[K, K]$ and a judicious expansion point to cover the eigenvalue domain of the problem.

3.2   Complex Z_2 noise trace estimation

Exact computation of the trace inverse for $N \times N$ matrices is very time consuming for matrices of size $N \sim 10^6$. However, the complex Z_2 noise method has been shown to provide an efficient stochastic estimation of the trace [4,6,7]. In fact, it has been proved to be an optimal choice for the noise, producing a *minimum* variance [8].

The complex Z_2 noise estimator can be briefly described as follows [4,8]. We construct L noise vectors $\eta^1, \eta^2, \cdots, \eta^L$ where $\eta^j = \{\eta_1^j, \eta_2^j, \eta_3^j, \cdots, \eta_N^j\}^T$, as follows. Each element η_n^j takes one of the four values $\{\pm 1, \pm\imath\}$ chosen independently with equal probability. It follows from the statistics of η_n^j that

$$E[< \eta_n >] \equiv E[\frac{1}{L} \sum_{j=1}^{L} \eta_n^j] = 0,$$

$$E[< \eta_m^\star \eta_n >] \equiv E[\frac{1}{L} \sum_{j=1}^{L} \eta_m^{\star j} \eta_n^j] = \delta_{mn}. \tag{11}$$

The vectors can be used to construct an unbiased estimator for the trace inverse of a given matrix M as follows:

$$E[< \eta^\dagger \mathbf{M}^{-1} \eta >] \equiv E[\frac{1}{L} \sum_{j=1}^{L} \sum_{m,n=1}^{N} \eta_m^{\star j} M_{m,n}^{-1} \eta_n^j]$$

$$= \sum_{n}^{N} M_{n,n}^{-1} + (\sum_{m \neq n}^{N} M_{m,n}^{-1})[\frac{1}{L} \sum_{j}^{L} \eta_m^{\star j} \eta_n^j]$$

$$= \text{Tr } \mathbf{M}^{-1}.$$

The variance of the estimator is shown to be [8]

$$\sigma_M^2 \equiv \text{Var}[< \eta^\dagger \mathbf{M}^{-1} \eta >] = \text{E}\left[| < \eta^\dagger \mathbf{M}^{-1} \eta > - \text{Tr } \mathbf{M}^{-1}|^2\right]$$

$$= \frac{1}{L} \sum_{m \neq n}^{N} M_{m,n}^{-1}(M_{m,n}^{-1})^\star = \frac{1}{L} \sum_{m \neq n}^{N} |M_{m,n}^{-1}|^2 .$$

The stochastic error of the complex Z_2 noise estimate results only from the off-diagonal entries of the inverse matrix (the same is true for Z_n noise for any n). However, other noises (such as Gaussian) have additional errors arising from diagonal entries. This is why the Z_2 noise has minimum variance. For example, it has been demonstrated on a $16^3 \times 24$ lattice with $\beta = 6.0$ and $\kappa = 0.148$ for the Wilson action that the Z_2 noise standard deviation is smaller than that of the Gaussian noise by a factor of 1.54 [4].

Applying the complex Z_2 estimator to the expression for the $Tr log M$ in Eq. (10), we find

$$\sum_k b_k \text{Tr}(M + c_k)^{-1}$$

$$\approx \frac{1}{L} \sum_k^K \sum_j^L b_k \eta^{j\dagger} (M + c_k)^{-1} \eta^j$$

$$= \frac{1}{L} \sum_j^L \sum_{k=1}^K b_k \eta^{j\dagger} \xi^{k,j}, \tag{12}$$

where $\xi^{k,j} = (M + c_k I)^{-1} \eta^j$ are the solutions of

$$(M + c_k I)\xi^{k,j} = \eta^j, \tag{13}$$

Since $M + c_k I$ are shifted matrices with constant diagonal matrix elements, Eq. (13) can be solved collectively for all values of c_k within one iterative process by several algorithms, including the Quasi-Minimum Residual (QMR) [9], Multiple-Mass Minimum Residual (M^3 R) [10], and GMRES[11]. We have adopted the M^3 R algorithm, which has been shown to be about 2 times faster than the conjugate gradient algorithm, and the overhead for the multiple c_k is only 8% [12]. The only price to pay is memory: for each c_k, a vector of the solution needs to be stored. Furthermore, one observes that $c_k > 0$. This improves the conditioning of $(M + c_k I)$ since the eigenvalues of M have positive real parts. Hence, we expect faster convergence for column inversions for Eq. (13).

In the next section, we describe a method which significantly reduces the stochastic error.

3.3   Improved PZ estimation with unbiased subtraction

In order to reduce the variance of the estimate, we introduce a suitably chosen set of traceless $N \times N$ matrices $\mathbf{Q}^{(p)}$, i.e. which satisfy $\sum_{n=1}^N \mathbf{Q}_{n,n}^{(p)} = 0$, $p = 1 \cdots P$. The expected value and variance for the modified estimator $< \eta^\dagger (\mathbf{M}^{-1} - \sum_{p=1}^P \lambda_p \mathbf{Q}^{(p)})\eta >$ are given by

$$E[< \eta^\dagger (\mathbf{M}^{-1} - \sum_{p=1}^P \lambda_p \mathbf{Q}^{(p)}))\eta >] = \text{Tr } \mathbf{M}^{-1}, \tag{14}$$

$$\Delta_M(\lambda) = \text{Var}[< \eta^\dagger (\mathbf{M}^{-1} - \sum_{p=1}^P \lambda_p \mathbf{Q}^{(p)})\eta >] \tag{15}$$

$$= \frac{1}{L} \sum_{m \neq n} |\mathbf{M}_{m,n}^{-1} - \sum_{p=1}^P \lambda_p \mathbf{Q}_{m,n}^{(p)}|^2,$$

for any values of the real parameters λ_p. In other words, introducing the matrices $\mathbf{Q}^{(p)}$ into the estimator produces no bias, but may reduce the error bars if the $\mathbf{Q}^{(p)}$ are chosen judiciously. Further, λ_p may be varied at will to achieve a minimum variance estimate: this corresponds to a least-squares fit to the function $\eta^\dagger \mathbf{M}^{-1}\eta$ sampled at points η_j, $j = 1 \cdots L$, using the fitting functions $\{1, \eta^\dagger \mathbf{Q}^{(p)}\eta\}$, $p = 1 \cdots P$.

We now turn to the question of choosing suitable traceless matrices $\mathbf{Q}^{(p)}$ to use in the modified estimator. One possibility for the Wilson fermion matrix $\mathbf{M} = \mathbf{I} - \kappa \mathbf{D}$ is suggested by the hopping parameter — κ expansion of the inverse matrix,

$$(\mathbf{M} + c_k \mathbf{I})^{-1} = \frac{1}{\mathbf{M} + c_k \mathbf{I}} = \frac{1}{(1 + c_k)(\mathbf{I} - \frac{\kappa}{(1+c_k)}\mathbf{D})} \qquad (16)$$

$$= \frac{\mathbf{I}}{1 + c_k} + \frac{\kappa}{(1 + c_k)^2}\mathbf{D} + \frac{\kappa^2}{(1 + c_k)^3}\mathbf{D}^2 + \frac{\kappa^3}{(1 + c_k)^4}\mathbf{D}^3 + \cdots .$$

This suggests choosing the matrices $\mathbf{Q}^{(p)}$ from among those matrices in the hopping parameter expansion which are traceless:

$$\mathbf{Q}^{(1)} = \frac{\kappa}{(1 + c_k)^2}\mathbf{D},$$

$$\mathbf{Q}^{(2)} = \frac{\kappa^2}{(1 + c_k)^3}\mathbf{D}^2,$$

$$\mathbf{Q}^{(3)} = \frac{\kappa^3}{(1 + c_k)^4}\mathbf{D}^3,$$

$$\mathbf{Q}^{(4)} = \frac{\kappa^4}{(1 + c_k)^5}(\mathbf{D}^4 - \mathrm{Tr}\mathbf{D}^4),$$

$$\mathbf{Q}^{(5)} = \frac{\kappa^5}{(1 + c_k)^6}\mathbf{D}^5,$$

$$\mathbf{Q}^{(6)} = \frac{\kappa^6}{(1 + c_k)^7}(\mathbf{D}^6 - \mathrm{Tr}\mathbf{D}^6),$$

$$\mathbf{Q}^{(2r+1)} = \frac{\kappa^{2r+1}}{(1 + c_k)^{2r+2}}\mathbf{D}^{2r+1}, \qquad r = 3, 4, 5, \cdots .$$

It may be verified that all of these matrices are traceless. In principle, one can include all the even powers which entails the explicit calculation of all the allowed loops in $Tr D^{2r}$. In this manuscript we have only included $\mathbf{Q}^{(4)}$, $\mathbf{Q}^{(6)}$, and $\mathbf{Q}^{(2r+1)}$.

3.4 Computation of $Tr \log M$

Our numerical computations were carried out with the Wilson action on the $8^3 \times 12$ ($N = 73728$) lattice with $\beta = 5.6$. We use the HMC with pseudofermions to generate gauge configurations. With a cold start, we obtain the fermion matrix M_1 after the plaquette becomes stable. The trajectories are traced with $\tau = 0.01$ and 30 molecular dynamics steps using $\kappa = 0.150$. M_2 is then obtained from M_1 by an accepted trajectory run. Hence M_1 and M_2 differ by a continuum perturbation, and $\log[\det M_1 / \det M_2] \sim \mathcal{O}(1)$.

We first calculate $\log \det M_1$ with different orders of Padé expansion around $z_0 = 0.1$ and $z_0 = 1.0$. We see from Table 2 that the 5th order Padé does not give the same answer for two different expansion points, suggesting that its accuracy is not sufficient for the range of eigenvalues of M_1. Whereas, the 11th order Padé gives the same answer within errors. Thus, we shall choose $P[11,11](z)$ with $z_0 = 0.1$ to perform the calculations from this point on.

Table 2. Unimproved and improved PZ estimates for $\log[\det M_1]$ with 100 complex Z_2 noise vectors. $\kappa = 0.150$.

$P[K,K](z)$	$K =$	5	7	9	11
$z_0 = 0.1$	Original:	473(10)	774(10)	796(10)	798(10)
	Improved:	487.25(62)	788.17(62)	810.83(62)	812.33(62)
$z_0 = 1.0$	Original:	798(10)	798(10)	798(10)	799(10)
	Improved:	812.60(62)	812.37(62)	812.36(62)	812.37(62)

In Table 3, we give the results of improved estimations for $\mathrm{Tr} \log M_1$. We see that the variational technique described above can reduce the data fluctuations by more than an order of magnitude. For example, the unimproved error $\delta_0 = 5.54$ in Table 3 for 400 Z_2 noises is reduced to $\delta_{11} = 0.15$ for the subtraction which includes up to the Q^{11} matrix. This is 37 times smaller. Comparing the central values in the last row (i.e. the 11th order improved) with that of unimproved estimate with 10,000 Z_2 noises, we see that they are the same within errors. This verifies that the variational subtraction scheme that we employed does not introduce biased errors. The improved estimates of $\mathrm{Tr} \log M_1$ from 50 Z_2 noises with errors δ_r from Table 3 are plotted in comparison with the central value of the unimproved estimate from 10,000 noises in Fig. 1.

4   Summary

In summary, the new noisy Monte Carlo algorithm proposed here is free from the problem of probability-bound violations which afflicts the linear accept/reject algorithm, especially when the variance of the noise is large. The upper-bound violation is avoided by going back to the Metropolis accept/reject. The lower-bound violation problem is tackled by grouping the sign of the estimated probability with the observable. With the probability-bound violation problem solved, NMC is a bona fide unbiased stochastic

Table 3. Central values for improved stochastic estimation of log[det M_1] and rth–order improved Jackknife errors δ_r are given for different numbers of Z_2 noise vectors. κ is 0.150 in this case.

# Z_2	50	100	200	400	600	800	1000	3000	10000
0th	802.98	797.98	810.97	816.78	815.89	813.10	816.53	813.15	812.81
δ_0	±14.0	±9.81	±7.95	±5.54	±4.47	±3.83	±3.41	±1.97	±1.08
1st	807.89	811.21	814.13	815.11	814.01	814.62	814.97	—	—
δ_1	±4.65	±3.28	±2.48	±1.84	±1.50	±1.29	±1.12	-	-
2nd	813.03	812.50	811.99	812.86	811.87	812.89	813.04	—	—
δ_2	±2.46	±1.65	±1.34	±1.01	±0.83	±0.72	±0.64	-	-
3rd	812.07	812.01	811.79	812.44	812.18	812.99	813.03	—	—
δ_3	±1.88	±1.31	±1.01	±0.74	±0.58	±0.51	±0.44	-	-
4th	812.28	812.52	812.57	812.86	812.85	813.25	813.40	—	—
δ_4	±1.20	±0.94	±0.68	±0.48	±0.39	±0.35	±0.30	-	-
5th	813.50	813.07	813.36	813.70	813.47	813.54	813.50	—	—
δ_5	±0.82	±0.62	±0.47	±0.34	±0.29	±0.25	±0.22	-	-
6^{ts}	813.54	813.23	813.22	813.28	813.20	813.37	813.26	—	—
δ_6	±0.67	±0.49	±0.35	±0.25	±0.21	±0.18	±0.16	-	-
7^{ts}	814.18	813.74	813.44	813.42	813.39	—	—	—	—
δ_7	±0.44	±0.36	±0.26	±0.19	±0.16	-	-	-	-
9th	813.77	813.62	813.49	813.40	813.43	—	—	—	—
δ_9	±0.40	±0.30	±0.22	±0.16	±0.14	-	-	-	-
11th	813.54	813.41	813.45	813.44	813.43	—	—	—	—
δ_{11}	±0.38	±0.27	±0.21	±0.15	±0.13	-	-	-	-

algorithm as demonstrated in the 5-state model. Furthermore, it is shown in the 5-state model that it is not necessary to have an extremely small variance in the stochastic estimation. With the encouraging results from the Padé-Z_2 estimation of the $\text{Tr} \ln M$ [5], one has a reasonable hope that the V^2 dependence of NMC will be tamed with a smaller prefactor. We will apply NMC to the dynamical fermion updating in QCD and compare it to the HMC with pseudo-fermions [13].

5   Acknowledgment

This work is partially supported by the U.S. DOE grant DE-FG05-84ER40154.

References

1. A.D. Kennedy, J. Kuti, Phys. Rev. Lett. **54** (1985) 2473.
2. L. Lin, K. F. Liu, and J. Sloan, hep-lat/9905033.
3. G. Bhanot, A. D. Kennedy, Phys. Lett. **157B** (1985) 70.
4. S. J. Dong and K. F. Liu, Phys. Lett. **B 328**, 130 (1994).

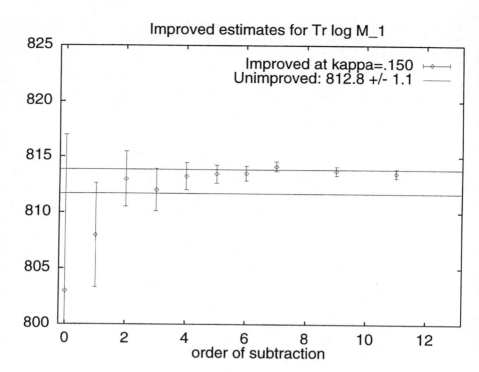

Fig. 1. The improved PZ estimate of $\mathrm{Tr}\log \mathbf{M}_1$ with 50 noises as a function of the order of subtraction and compared to that of unimproved estimate with 10,000 noises. The dashed lines are drawn with a distance of 1 σ away from the central value of the unimproved estimate.

5. C. Thron, S. J. Dong, K. F. Liu, H. P. Ying, Phys. Rev. **D57** (1998) 1642.
6. S.J. Dong, J.-F. Lagaë, and K.F. Liu, Phys. Rev. Lett. **75**, 2096 (1995); S.J. Dong, J.-F. Lagaë, and K.F. Liu, Phys. Rev. **D54**, 5496 (1996).
7. N. Eicker, *et al*, SESAM-Collaboration, Phys. Lett. **B389**, 720 (1996).
8. S. Bernardson, P. McCarty and C. Thron, Comp. Phys. Commun. **78**, 256 (1994).
9. A. Frommer, B. Nöckel, S. Güsken, Th. Lippert and K. Schilling, Int. J. Mod. Phys. **C6**, 627 (1995).
10. U. Glässner, S. Güsken, Th. Lippert, G. Ritzenhöfer, K. Schilling, and A. Frommer. How to compute Green's functions for entire mass trajectories within Krylov solvers. *Int. J. Mod. Phys.*, C7:635, 1996.
11. A. Frommer and U. Glässner, Wuppertal preprint BUGHW-SC96/8, in SIAM J. Scientific Computing.
12. He-Ping Ying, S.J. Dong and K.F. Liu, Nucl. Phys. **B**(proc. Suppl.) **53**, 993 (1997).
13. S. Duane, A. D. Kennedy, B. J. Pendleton, D. Roweth, Phys. Lett. **B195** (1987) 216.

Least-Squares Optimized Polynomials for Fermion Simulations

István Montvay

Deutsches Elektronen-Synchrotron DESY
Notkestr. 85
D-22603 Hamburg, Germany

Abstract. Least-squares optimized polynomials are discussed which are needed in the two-step multi-bosonic algorithm for Monte Carlo simulations of quantum field theories with fermions. A recurrence scheme for the calculation of necessary coefficients in the recursion and for the evaluation of these polynomials is introduced.

1 Introduction

In popular Monte Carlo simulation algorithms for QCD and other similar quantum field theories the main difficulty is the evaluation of the determinant of the fermion action matrix. This can be achieved by stochastic procedures with the help of auxiliary bosonic "pseudofermion" fields.

In the *two-step multi-bosonic (TSMB) algorithm* [1] an approximation of the fermion determinant is achieved by the pseudofermion fields corresponding to a polynomial approximation of some negative power $x^{-\alpha}$ of the fermion matrix [2]. The auxiliary bosonic fields are updated according to the *multi-bosonic action* [3]. The error of the polynomial approximation is corrected in a global accept-reject decision by using better polynomial approximations. Sometimes a reweighting of gauge configurations in the evaluation of expectation values is also necessary. This can also be performed by high order polynomials.

The polynomials used in the TSMB algorithm have to approximate the function $x^{-\alpha}\bar{P}(x)$ in some non-negative interval $x \in [\epsilon, \lambda]$, $0 \leq \epsilon < \lambda$. Here $\bar{P}(x)$ is a known polynomial, typically another cruder approximation of $x^{-\alpha}$. The approximation scheme and optimization procedure can be chosen differently. The least-squares optimization [4] is an efficient and flexible possibility. (For other approximation schemes see [5,6].)

In this review the basic relations for least-squares optimized polynomials are presented as introduced in [1,2]. Particular attention is paid to a recurrence scheme which can be applied for determining the necessary high order polynomials and for evaluating them numerically. The details of the TSMB algorithm will not be considered. For a comprehensive summary and references see [7]. For experience on the application of TSMB in a recent large scale numerical simulation see [8,9].

2   Basic Relations

Least-squares optimization provides a general and flexible framework for obtaining the necessary optimized polynomials in multi-bosonic fermion algorithms. Here we introduce the basic formulae in the way it has been done in [2,9].

We want to approximate the real function $f(x)$ in the interval $x \in [\epsilon, \lambda]$ by a polynomial $P_n(x)$ of degree n. The aim is to minimize the deviation norm

$$\delta_n \equiv \left\{ N_{\epsilon,\lambda}^{-1} \int_\epsilon^\lambda dx\, w(x)^2 \left[f(x) - P_n(x) \right]^2 \right\}^{\frac{1}{2}} . \tag{1}$$

Here $w(x)$ is an arbitrary real weight function and the overall normalization factor $N_{\epsilon,\lambda}$ can be chosen by convenience, for instance, as

$$N_{\epsilon,\lambda} \equiv \int_\epsilon^\lambda dx\, w(x)^2 f(x)^2 . \tag{2}$$

A typical example of functions to be approximated is $f(x) = x^{-\alpha}/\bar{P}(x)$ with $\alpha > 0$ and some polynomial $\bar{P}(x)$. The interval is usually such that $0 \le \epsilon < \lambda$. For optimizing the relative deviation one takes a weight function $w(x) = f(x)^{-1}$.

δ_n^2 is a quadratic form in the coefficients of the polynomial which can be straightforwardly minimized. Let us now consider, for simplicity, only the relative deviation from the simple function $f(x) = x^{-\alpha} = w(x)^{-1}$. Let us denote the polynomial corresponding to the minimum of δ_n by

$$P_n(\alpha; \epsilon, \lambda; x) \equiv \sum_{\nu=0}^n c_{n\nu}(\alpha; \epsilon, \lambda) x^{n-\nu} . \tag{3}$$

Performing the integral in δ_n^2 term by term we obtain

$$\delta_n^2 = 1 - 2 \sum_{\nu=0}^n c_\nu V_\nu^{(\alpha)} + \sum_{\nu_1,\nu_2=0}^n c_{\nu_1} M_{\nu_1,\nu_2}^{(\alpha)} c_{\nu_2} , \tag{4}$$

where

$$V_\nu^{(\alpha)} = \frac{\lambda^{1+\alpha+n-\nu} - \epsilon^{1+\alpha+n-\nu}}{(\lambda - \epsilon)(1 + \alpha + n - \nu)} ,$$

$$M_{\nu_1,\nu_2}^{(\alpha)} = \frac{\lambda^{1+2\alpha+2n-\nu_1-\nu_2} - \epsilon^{1+2\alpha+2n-\nu_1-\nu_2}}{(\lambda - \epsilon)(1 + 2\alpha + 2n - \nu_1 - \nu_2)} . \tag{5}$$

The coefficients of the polynomial corresponding to the minimum of δ_n^2, or of δ_n, are

$$c_\nu \equiv c_{n\nu}(\alpha; \epsilon, \lambda) = \sum_{\nu_1=0}^n M_{\nu\nu_1}^{(\alpha)-1} V_{\nu_1}^{(\alpha)} . \tag{6}$$

The value at the minimum is

$$\delta_n^2 \equiv \delta_n^2(\alpha; \epsilon, \lambda) = 1 - \sum_{\nu_1, \nu_2 = 0}^{n} V_{\nu_1}^{(\alpha)} M_{\nu_1, \nu_2}^{(\alpha)-1} V_{\nu_2}^{(\alpha)} . \tag{7}$$

The solution of the quadratic optimization in (6)-(7) gives in principle a simple way to find the required least-squares optimized polynomials. The practical problem is, however, that the matrix $M^{(\alpha)}$ is not well conditioned because it has eigenvalues with very different magnitudes. In order to illustrate this let us consider the special case ($\alpha = 1$, $\lambda = 1$, $\epsilon = 0$) with $n = 10$. In this case the eigenvalues are:

$$0.4435021205e - 14 , \quad 0.1045947635e - 11 , \quad 0.1143819915e - 9 ,$$

$$0.7698917100e - 8 , \quad 0.3571195735e - 6 , \quad 0.1211873623e - 4 ,$$

$$0.3120413130e - 3 , \quad 0.6249495675e - 2 , \quad 0.9849331094e - 1 ,$$

$$1.075807246 . \tag{8}$$

A numerical investigation shows that, in general, the ratio of maximal to minimal eigenvalues is of the order of $\mathcal{O}(10^{1.5n})$. It is obvious from the structure of $M^{(\alpha)}$ in (5) that a rescaling of the interval $[\epsilon, \lambda]$ does not help. The large differences in magnitude of the eigenvalues implies through (6) large differences of magnitude in the coefficients $c_{n\nu}$ and therefore the numerical evaluation of the optimal polynomial $P_n(x)$ for large n is non-trivial.

Let us now return to the general case with arbitrary function $f(x)$ and weight $w(x)$. It is very useful to introduce orthogonal polynomials $\Phi_\mu(x)$ ($\mu = 0, 1, 2, \dots$) satisfying

$$\int_\epsilon^\lambda dx\, w(x)^2 \Phi_\mu(x) \Phi_\nu(x) = \delta_{\mu\nu} q_\nu . \tag{9}$$

and expand the polynomial $P_n(x)$ in terms of them:

$$P_n(x) = \sum_{\nu=0}^{n} d_{n\nu} \Phi_\nu(x) . \tag{10}$$

Besides the normalization factor q_ν let us also introduce, for later purposes, the integrals p_ν and s_ν by

$$q_\nu \equiv \int_\epsilon^\lambda dx\, w(x)^2 \Phi_\nu(x)^2 ,$$

$$p_\nu \equiv \int_\epsilon^\lambda dx\, w(x)^2 \Phi_\nu(x)^2 x ,$$

$$s_\nu \equiv \int_\epsilon^\lambda dx\, w(x)^2 x^\nu . \tag{11}$$

It can be easily shown that the expansion coefficients $d_{n\nu}$ minimizing δ_n are independent of n and are given by

$$d_{n\nu} \equiv d_\nu = \frac{b_\nu}{q_\nu} , \tag{12}$$

where

$$b_\nu \equiv \int_\epsilon^\lambda dx\, w(x)^2 f(x) \Phi_\nu(x) . \tag{13}$$

The minimal value of δ_n^2 is

$$\delta_n^2 = 1 - N_{\epsilon,\lambda}^{-1} \sum_{\nu=0}^n d_\nu b_\nu . \tag{14}$$

Rescaling the variable x by $x' = \rho x$ allows for considering only standard intervals, say $[\epsilon/\lambda, 1]$. The scaling properties of the optimized polynomials can be easily obtained from the definitions. Let us now again consider the simple function $f(x) = x^{-\alpha}$ and relative deviation with $w(x) = x^\alpha$ when the rescaling relations are:

$$\delta_n^2(\alpha; \epsilon\rho, \lambda\rho) = \delta_n^2(\alpha; \epsilon, \lambda) ,$$

$$P_n(\alpha; \epsilon\rho, \lambda\rho; x) = \rho^{-\alpha} P_n(\alpha; \epsilon, \lambda; x/\rho) ,$$

$$c_{n\nu}(\alpha; \epsilon\rho, \lambda\rho) = \rho^{\nu-n-\alpha} c_{n\nu}(\alpha; \epsilon, \lambda) . \tag{15}$$

In applications to multi-bosonic algorithms for fermions the decomposition of the optimized polynomials as a product of root-factors is needed. This can be written as

$$P_n(\alpha; \epsilon, \lambda; x) = c_{n0}(\alpha; \epsilon, \lambda) \prod_{j=1}^n [x - r_{nj}(\alpha; \epsilon, \lambda)] . \tag{16}$$

The rescaling properties here are:

$$c_{n0}(\alpha; \epsilon\rho, \lambda\rho) = \rho^{-n-\alpha} c_{n0}(\alpha; \epsilon, \lambda) ,$$

$$r_{nj}(\alpha; \epsilon\rho, \lambda\rho) = \rho\, r_{nj}(\alpha; \epsilon, \lambda) . \tag{17}$$

The root-factorized form (16) can also be used for the numerical evaluation of the polynomials with matrix arguments if a suitable optimization of the ordering of roots is performed [2].

The above orthogonal polynomials satisfy three-term recurrence relations which are very useful for numerical evaluation. In fact, at large n the recursive evaluation of the polynomials is numerically more stable than the evaluation with root factors. For general $f(x)$ and $w(x)$, the first two orthogonal polynomials with $\mu = 0, 1$ are given by

$$\Phi_0(x) = 1 , \qquad \Phi_1(x) = x - \frac{s_1}{s_0} . \tag{18}$$

The higher order polynomials $\Phi_\mu(x)$ for $\mu = 2, 3, \ldots$ can be obtained from the recurrence relation

$$\Phi_{\mu+1}(x) = (x + \beta_\mu)\Phi_\mu(x) + \gamma_{\mu-1}\Phi_{\mu-1}(x) , \qquad (\mu = 1, 2, \ldots) , \tag{19}$$

where the recurrence coefficients are given by

$$\beta_\mu = -\frac{p_\mu}{q_\mu} , \qquad \gamma_{\mu-1} = -\frac{q_\mu}{q_{\mu-1}} . \tag{20}$$

Defining the polynomial coefficients $f_{\mu\nu}$ $(0 \le \nu \le \mu)$ by

$$\Phi_\mu(x) = \sum_{\nu=0}^{\mu} f_{\mu\nu} x^{\mu-\nu} \tag{21}$$

the above recurrence relations imply the normalization convention

$$f_{\mu 0} = 1 , \qquad (\mu = 0, 1, 2, \ldots) . \tag{22}$$

The rescaling relations for the orthogonal polynomials easily follow from the definitions. For the simple function $f(x) = x^{-\alpha}$ and relative deviation with $w(x) = x^\alpha$ we have

$$\Phi_\mu(\alpha; \rho\epsilon, \rho\lambda; x) = \rho^\mu \Phi_\mu(\alpha; \epsilon, \lambda; x/\rho) . \tag{23}$$

For the quantities introduced in (11) this implies

$$q_\nu(\alpha; \rho\epsilon, \rho\lambda) = \rho^{2\alpha+1+2\nu} q_\nu(\alpha; \epsilon, \lambda) ,$$

$$p_\nu(\alpha; \rho\epsilon, \rho\lambda) = \rho^{2\alpha+2+2\nu} p_\nu(\alpha; \epsilon, \lambda) ,$$

$$s_\nu(\alpha; \rho\epsilon, \rho\lambda) = \rho^{2\alpha+1+\nu} s_\nu(\alpha; \epsilon, \lambda) . \tag{24}$$

For the expansion coefficients defined in (12)-(13) one obtains

$$b_\nu(\alpha; \rho\epsilon, \rho\lambda) = \rho^{\alpha+1+\nu} b_\nu(\alpha; \epsilon, \lambda) ,$$

$$d_\nu(\alpha; \rho\epsilon, \rho\lambda) = \rho^{-\alpha-\nu} d_\nu(\alpha; \epsilon, \lambda) , \tag{25}$$

and the recurrence coefficients in (19)-(20) satisfy

$$\beta_\mu(\alpha; \rho\epsilon, \rho\lambda) = \rho\, \beta_\mu(\alpha; \epsilon, \lambda) ,$$

$$\gamma_{\mu-1}(\alpha; \rho\epsilon, \rho\lambda) = \rho^2\, \gamma_{\mu-1}(\alpha; \epsilon, \lambda) . \tag{26}$$

For general intervals $[\epsilon, \lambda]$ and/or functions $f(x) = x^{-\alpha}\bar{P}(x)$ the orthogonal polynomials and expansion coefficients have to be determined numerically. In some special cases, however, the polynomials can be related to some well known ones. An example is the weight factor

$$w^{(\rho,\sigma)}(x)^2 = (x - \epsilon)^\rho(\lambda - x)^\sigma .$$ (27)

Taking, for instance, $\rho = 2\alpha$, $\sigma = 0$ this weight is similar to the one for relative deviation from the function $f(x) = x^{-\alpha}$, which would be just $x^{2\alpha}$. In fact, for $\epsilon = 0$ these are exactly the same and for small ϵ the difference is negligible. The corresponding orthogonal polynomials are simply related to the Jacobi polynomials [9], namely

$$\Phi_\nu^{(\rho,\sigma)}(x) = (\lambda - \epsilon)^\nu \nu! \frac{\Gamma(\rho + \sigma + \nu + 1)}{\Gamma(\rho + \sigma + 2\nu + 1)} P_\nu^{(\sigma,\rho)}\left(\frac{2x - \lambda - \epsilon}{\lambda - \epsilon}\right) .$$ (28)

Comparing different approximations with different (ρ, σ) the best choice is usually $\rho = 2\alpha$, $\sigma = 0$ which corresponds to optimizing the relative deviation (see the appendix of [9]).

For large condition numbers λ/ϵ least-squares optimization is much better than the Chebyshev approximation used for the approximation of x^{-1} in [3]. The Chebyshev approximation is minimizing the maximum of the relative deviation

$$R(x) \equiv xP(x) - 1 .$$ (29)

For the deviation norm

$$\delta_{max} \equiv \max_{x \in [\epsilon, \lambda]} |R(x)|$$ (30)

the least-squares approximation is slightly worse than the Chebyshev approximation. An example is shown by fig. 1. In the left lower corner the Chebyshev approximation has $Rc(0.0002) = -0.968$ compared to $Ro(0.0002) = -0.991$ for the least-squares optimization. For smaller condition numbers the Chebyshev approximation is not as bad as is shown by fig. 1. Nevertheless, in QCD simulations in sufficiently large volumes the condition number is of the order of the light quark mass squared in lattice units which can be as large as $\mathcal{O}(10^6 - 10^7)$.

Figure 1 also shows that the least-squares optimization is quite good in the minimax norm in (30), too. It can be proven that

$$|Ro(\epsilon)| = \max_{x \in [\epsilon, \lambda]} |Ro(x)|$$ (31)

hence the minimax norm can also be easily obtained from

$$\delta_{max}^{(o)} = \max_{x \in [\epsilon, \lambda]} |Ro(x)| = |Ro(\epsilon)| .$$ (32)

relative deviation from 1/x: n=16, in [0.0002,3.5]

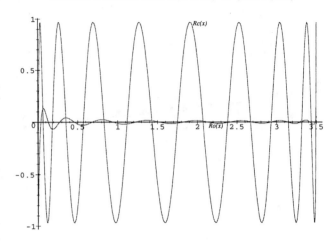

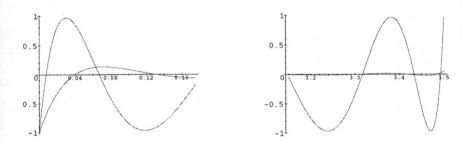

Fig. 1. The comparison of the relative deviation $R(x) \equiv xP(x) - 1$ for the Chebyshev polynomial $(Rc(x))$ and the quadratically optimized polynomial $(Ro(x))$. In the lower part the two ends of the interval are zoomed.

Therefore the least squares-optimization is also well suited for controlling the minimax norm, if for some reason it is required.

In QCD simulations the inverse power to be approximated, α, is related to the number of Dirac fermion flavours: $\alpha = N_f/2$. If only u- and d-quarks are considered we have $N_f = 2$ and the function to be approximated is x^{-1}.

The dependence of the (squared) least-squares norm in (1) on the polynomial order n is shown by fig. 2 for different values of the condition number λ/ϵ. The dependence on $\alpha = N_f/2$ is illustrated by fig. 3.

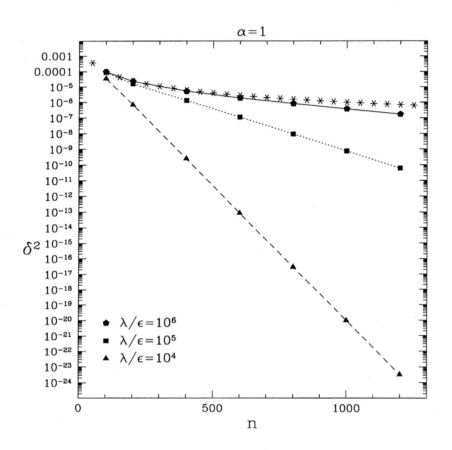

Fig. 2. The (squared) deviation norm δ^2 of the polynomial approximations of x^{-1} as function of the order for different values of λ/ϵ. The asterisks show the $\epsilon/\lambda \to 0$ limit.

Another possible application of least-squares optimized polynomials is the numerical evaluation of the zero mass lattice action proposed by Neuberger [10]. If one takes, for instance, the weight factor in (27) corresponding to the relative deviation, then the function $x^{-1/2}$ has to be expanded in the Jacobi polynomials $P^{(1,0)}$.

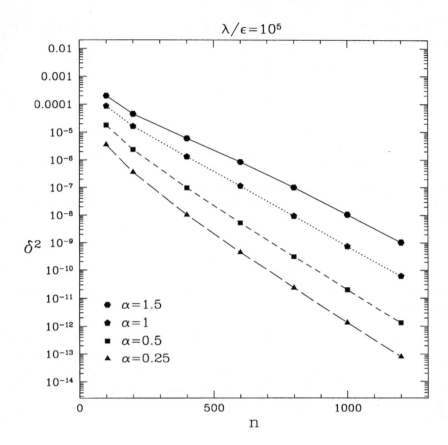

Fig. 3. The (squared) deviation norm δ^2 of the polynomial approximations of $x^{-\alpha}$ as function of the order at $\lambda/\epsilon = 10^5$ for different values of α.

3   Recurrence Scheme

The expansion in orthogonal polynomials is very useful because it allows for a numerically stable evaluation of the least-squares optimized polynomials by the recurrence relation (19). The orthogonal polynomials themselves can also be determined recursively.

A recurrence scheme for obtaining the recurrence coefficients $\beta_\mu, \gamma_{\mu-1}$ and expansion coefficients $d_\nu = b_\nu/q_\nu$ has been given in [8,7]. In order to obtain q_ν, p_ν contained in (20) one can use the relations

$$q_\mu = \sum_{\nu=0}^{\mu} f_{\mu\nu} s_{2\mu-\nu} \ , \qquad p_\mu = \sum_{\nu=0}^{\mu} f_{\mu\nu} \left(s_{2\mu+1-\nu} + f_{\mu 1} s_{2\mu-\nu} \right) \ . \qquad (33)$$

The coefficients themselves can be calculated from $f_{11} = -s_1/s_0$ and (19) which gives

$$
\begin{aligned}
f_{\mu+1,1} &= f_{\mu,1} + \beta_\mu \ , \\
f_{\mu+1,2} &= f_{\mu,2} + \beta_\mu f_{\mu,1} + \gamma_{\mu-1} \ , \\
f_{\mu+1,3} &= f_{\mu,3} + \beta_\mu f_{\mu,2} + \gamma_{\mu-1} f_{\mu-1,1} \ , \\
&\quad \cdots \\
f_{\mu+1,\mu} &= f_{\mu,\mu} + \beta_\mu f_{\mu,\mu-1} + \gamma_{\mu-1} f_{\mu-1,\mu-2} \ , \\
f_{\mu+1,\mu+1} &= \beta_\mu f_{\mu,\mu} + \gamma_{\mu-1} f_{\mu-1,\mu-1} \ .
\end{aligned}
\tag{34}
$$

The orthogonal polynomial and recurrence coefficients are recursively determined by (20) and (33)-(34). The expansion coefficients for the optimized polynomial $P_n(x)$ can be obtained from

$$
b_\mu = \sum_{\nu=0}^{\mu} f_{\mu\nu} \int_\epsilon^\lambda dx \, w(x)^2 f(x) x^{\mu-\nu} \ .
\tag{35}
$$

The ingredients needed for this recursion are the basic integrals s_ν defined in (11) and

$$
t_\nu \equiv \int_\epsilon^\lambda dx \, w(x)^2 f(x) x^\nu \ .
\tag{36}
$$

The recurrence scheme based on the coefficients of the orthogonal polynomials $f_{\mu\nu}$ in (34) is not optimal for large orders n, neither for arithmetics nor for storage requirements. A better scheme can be built up on the basis of the integrals

$$
r_{\mu\nu} \equiv \int_\epsilon^\lambda dx \, w(x)^2 \Phi_\mu(x) x^\nu \ ,
$$

$$
\mu = 0, 1, \ldots, n; \quad \nu = \mu, \mu+1, \ldots, 2n - \mu \ .
\tag{37}
$$

The recurrence coefficients β_μ, $\gamma_{\mu-1}$ can be expressed from

$$
q_\mu = r_{\mu\mu} \ , \qquad p_\mu = r_{\mu,\mu+1} + f_{\mu 1} r_{\mu\mu}
\tag{38}
$$

and eq. (20) as

$$
\beta_\mu = -f_{\mu 1} - \frac{r_{\mu,\mu+1}}{r_{\mu\mu}} \ , \qquad \gamma_{\mu-1} = -\frac{r_{\mu\mu}}{r_{\mu-1,\mu-1}} \ .
\tag{39}
$$

It follows from the definition that

$$
r_{0\nu} = \int_\epsilon^\lambda dx \, w(x)^2 x^\nu = s_\nu \ ,
$$

$$
r_{1\nu} = \int_\epsilon^\lambda dx \, w(x)^2 (x^{\nu+1} + f_{11} x^\nu) = s_{\nu+1} + f_{11} s_\nu \ .
\tag{40}
$$

The recurrence relation (19) for the orthogonal polynomials implies

$$r_{\mu+1,\nu} = r_{\mu,\nu+1} + \beta_\mu r_{\mu\nu} + \gamma_{\mu-1} r_{\mu-1,\nu} \ . \tag{41}$$

This has to be supplemented by

$$f_{11} = -\frac{s_1}{s_0} \tag{42}$$

and by the first equation from (34):

$$f_{\mu+1,1} = f_{\mu,1} + \beta_\mu \ . \tag{43}$$

Eqs. (39)-(43) define a complete recurrence scheme for determining the orthogonal polynomials $\Phi_\mu(x)$. The moments s_ν of the integration measure defined in (11) serve as the basic input in this scheme.

The integrals b_ν in (13), which are necessary for the expansion coefficients d_ν in (12), can also be calculated in a similar scheme built up on the integrals

$$b_{\mu\nu} \equiv \int_\epsilon^\lambda dx\, w(x)^2 f(x) \Phi_\mu(x) x^\nu \ ,$$

$$\mu = 0, 1, \ldots, n; \qquad \nu = 0, 1, \ldots, n - \mu \ . \tag{44}$$

The relations corresponding to (40)-(41) are now

$$b_{0\nu} = \int_\epsilon^\lambda dx\, w(x)^2 f(x) x^\nu = t_\nu \ ,$$

$$b_{1\nu} = \int_\epsilon^\lambda dx\, w(x)^2 f(x)(x^{\nu+1} + f_{11} x^\nu) = t_{\nu+1} + f_{11} t_\nu \ ,$$

$$b_{\mu+1,\nu} = b_{\mu,\nu+1} + \beta_\mu b_{\mu\nu} + \gamma_{\mu-1} b_{\mu-1,\nu} \ . \tag{45}$$

The only difference compared to (40)-(41) is that the moments of $w(x)^2$ are now replaced by the ones of $w(x)^2 f(x)$.

It is interesting to collect the quantities which have to be stored in order that the recurrence can be resumed. This is useful if after stopping the iterations, for some reason, the recurrence will be restarted. Let us assume that the quantities q_ν ($\nu = 0, \ldots, n$), b_ν ($\nu = 0, \ldots, n$), β_ν ($\nu = 1, \ldots, n-1$) and γ_ν ($\nu = 0, \ldots, n-2$) are already known and one wants to resume the recurrence in order to calculate these quantities for higher indices. For this it is enough to know the values of

$$f_{n-1,1} \ , \quad r_{n-1,n-1} \ ,$$

$$R_{0\ldots n}^{(0)} \equiv (r_{0,2n+1}, r_{1,2n}, \ldots, r_{n,n+1}) \ ,$$

$$R_{0\ldots n}^{(1)} \equiv (r_{0,2n}, r_{1,2n-1}, \ldots, r_{n,n}) \ ,$$

$$B_{0\ldots n}^{(1)} \equiv (b_{0,n}, b_{1,n-1}, \ldots, b_{n,0}) \ ,$$

$$B_{0\ldots n-1}^{(2)} \equiv (b_{0,n-1}, b_{1,n-2}, \ldots, b_{n-1,0}) \ . \tag{46}$$

This shows that for maintaining a *resumable recurrence* it is enough to store a set of quantities linearly increasing in n.

An interesting question is the increase of computational load as a function of the highest required order n. At the first sight this seems to go just like n^2, which is surprising because, as eq. (6) shows, finding the minimum requires the inversion of an $n \otimes n$ matrix. However, numerical experience shows that the number of required digits for obtaining a precise result does also increase linearly with n. This is due to the linearly increasing logarithmic range of eigenvalues, as illustrated by (8). Using, for instance, Maple V for the arbitrary precision arithmetic, the computation slows down by another factor going roughly as (but somewhat slower than) n^2. Therefore, the total slowing down in n is proportional to n^4. For the same reason the storage requirements increase by n^2.

4   A Convenient Choice for TSMB

In the TSMB algorithm for Monte Carlo simulations of fermionic theories, besides the simple function $x^{-\alpha}$, also the function $x^{-\alpha}/\bar{P}(x)$ has to be approximated. Here $\bar{P}(x)$ is typically a lower order approximation to $x^{-\alpha}$. In this case, if one chooses to optimize the relative deviation, the basic integrals defined in (11) and (36) are, respectively,

$$s_\nu = \int_\epsilon^\lambda dx \, \bar{P}(x)^2 x^{2\alpha+\nu} \ ,$$

$$t_\nu = \int_\epsilon^\lambda dx \, \bar{P}(x) x^{\alpha+\nu} \ . \tag{47}$$

It is obvious that, if the recurrence coefficients for the expansion of the polynomial $\bar{P}(x)$ in orthogonal polynomials are known, the recursion scheme can also be used for the evaluation of s_ν and t_ν.

Another observation is that the integrals in (47) can be simplified if, instead of choosing the weight factor $w(x)^2 = \bar{P}(x)^2 x^{2\alpha}$, one takes

$$w(x)^2 = \bar{P}(x) x^\alpha \ , \tag{48}$$

which leads to

$$s_\nu = \int_\epsilon^\lambda dx \, \bar{P}(x) x^{\alpha+\nu} \ ,$$

$$t_\nu = \int_\epsilon^\lambda dx \, x^\nu \ . \tag{49}$$

Since $\bar{P}(x)$ is an approximation to $x^{-\alpha}$, the function $f(x) \equiv x^{-\alpha}/\bar{P}(x)$ is close to one and the difference between $f(x)^{-2}$ and $f(x)^{-1}$ is small. Therefore the least-squares optimized approximations with the weights $w(x)^2 = f(x)^{-2}$ and $w(x)^2 = f(x)^{-1}$ are also similar. It turns out that the second choice is, in fact, a little bit better because the largest deviation from $x^{-\alpha}$ typically occurs at the lower end of the interval $x = \epsilon$ where $\bar{P}(x)$ is smaller than $x^{-\alpha}$. As a consequence, $\bar{P}(x)x^\alpha < 1$ and $\bar{P}(x)x^\alpha > \bar{P}(x)^2 x^{2\alpha}$. This means that choosing the weight factor in (48) is emphasising more the lower end of the interval where $\bar{P}(x)$ as an approximation of $x^{-\alpha}$ is worst.

In summary: least-squares optimization is a flexible and powerful tool which can serve as a basis for applying the two-step multi-bosonic algorithm for Monte Carlo simulations of QCD and other similar theories. With the help of the recurrence scheme described in the previous section one can determine the necessary polynomial approximations to high enough orders.

References

1. Montvay, I.: An algorithm for gluinos on the lattice. Nucl. Phys. **B466** (1996) 259–284
2. Montvay, I.: Quadratically optimized polynomials for fermion simulations. Comput. Phys. Commun. **109** (1998) 144–160
3. Lüscher, M.: A new approach to the problem of dynamical quarks in numerical simulations of lattice QCD. Nucl. Phys. **B418** (1994) 637–648
4. Rivlin, T.J.: *An Introduction to the Approximation of Functions*, Blaisdell Publ. Company, 1969.
5. Wolff, U.: Multiboson simulation of the Schrödinger functional. Nucl. Phys. Proc. Suppl. **63** (1998) 937-939
6. de Forcrand, P.: UV-filtered fermionic Monte Carlo. Nucl. Phys. Proc. Suppl. **73** (1999) 822-824.
7. Montvay, I.: Multi-bosonic algorithms for dynamical fermion simulations. Workshop on Molecular Dynamics on Parallel Computers, Jülich, Germany, February 1999, to appear in the proceedings; hep-lat/9903029.
8. Kirchner, R. et al.: Evidence for discrete chiral symmetry breaking in N=1 supersymmetric Yang-Mills theory. Phys. Lett. **B446** (1999) 209–215
9. Campos, I. et al.: Monte Carlo simulation of SU(2) Yang-Mills theory with light gluinos. Eur. Phys. J. C; DOI 10.1007/s100529900183.
10. Neuberger, H.: Exactly massless quarks on the lattice. Phys. Lett. **B417** (1998) 141–144.

One-Flavour Hybrid Monte Carlo with Wilson Fermions

Thomas Lippert

University of Wuppertal
Department of Physics
D-42097 Wuppertal, Germany

Abstract. The Wilson fermion determinant can be written as product of the determinants of two hermitian positive definite matrices. This formulation allows to simulate non-degenerate quark flavors by means of the hybrid Monte Carlo algorithm. A major numerical difficulty is the occurrence of nested inversions. We construct a Uzawa iteration scheme which treats the nested system within one iterative process.

1 Introduction

The strong interaction between the quarks is described by quantum chromodynamics (QCD), a constitutive part of the standard model of elementary particle physics. QCD develops a strong coupling constant at the hadronic mass scale and thus, perturbation theory cannot be applied. The only known non-perturbative ab-initio method is to simulate QCD on a 4-dimensional space-time lattice by use of Monte Carlo methods as known from statistical physics.

One would think that lattice simulations of QCD have to take into account virtual loops from all six quark flavors, up (u), down (d), strange (s), charm (c) bottom (b), and top (t). Their masses span a wide range from about 3 MeV to about 180 GeV. [1] The masses of the three heavy quarks lie above the momentum scale set by dynamical chiral symmetry breaking. The QCD Lagrangian is chirally symmetric for vanishing quark masses. Chiral symmetry is explicitly broken by the quark masses and spontaneously by the dynamics. For light quarks, (u, d, s), dynamical breaking dominates, as signaled by zero mass Goldstone bosons, i.e., the pion octet. Supposedly only the light quarks contribute with virtual loops; the contributions of the heavier ones are assumed to be negligible.

A straightforward lattice QCD simulation including u, d, and s is difficult: on one hand, u and d are very light. As we know from chiral perturbation

---

[1] The light quark (u, d, s) masses are so-called "current" masses determined within the $\overline{\text{MS}}$ scheme at a renormalization scale $\mu = 2$ GeV. The masses of the heavy quarks (c, b, t) are "running" masses determined at $\mu = m_{c,b,t}$ in the $\overline{\text{MS}}$ scheme [1].

theory, the square of the pion mass is proportional to the quark mass [2]. Therefore, the pion acquires a small mass, i.e. a large correlation length, $\xi = \frac{1}{ma}$, for u and d masses approaching the chiral limit, m_u and $m_d = 0$. a denotes the lattice spacing. The lattice volume must be larger by more than one order of magnitude than possible today to reach physical parameter values for m_u and m_d. Hence, one has to recourse to extrapolations using results of simulations at artificial parameters for m_u and m_d and small lattices, far off the chiral limit. These difficulties go along with an increasing condition number of the fermionic matrix M, therefore its inversion suffers from critical slowing down approaching the chiral limit [3,4].

On the other hand, until recently, the only exact simulation algorithm for QCD with dynamical fermions[2] was the hybrid Monte Carlo algorithm (HMC) [5,6]. The benefits of HMC in reducing the computational complexity are achieved by treating the fermionic determinants as a stochastic estimate using Gaussian random fields [7], with the advantage, that instead of computing the determinant, only the solution of a linear system is required. For this approach the fermion matrix M must be hermitian positive definite (*hpd*).

The Wilson fermion discretization of the fermionic sector of QCD describes single flavors [8], in contrast to staggered fermions that represent four fermions, intermixed in spin-flavor space.[3] Unfortunately, Wilson fermions must violate chiral symmetry, see Ref. [9]. As a consequence, the Wilson fermion matrix is complex non-normal and thus cannot be included in the HMC scheme. To avoid this problem, one usually simulates two mass degenerate light quarks, an approximation justified by the fact that both quarks, u and d, are light and close in mass compared to the next heavier one, s. The product of two mass degenerate determinants indeed amounts to an *hpd* matrix, that can be simulated by the HMC. The minor price to pay is that the two light quarks are mass-degenerate, the major price is that the s quark is not included in the simulation.

There are attempts to evaluate operators, that contain the s quark, by application of a "partially quenched analysis" [10]. The PQA extrapolates only one of the valence quarks towards the chiral limit and holds the other one at the mass of the s quark—of course within the light mass-degenerate u-d-sea. PQA, however, leads to inconsistencies: the so-called J-parameter does not acquire its physical value [11,12].

One can try to employ an approximate one-flavor determinant representation. Such simulations, however, are plagued by systematic uncertainties, as one has to recourse to non-exact simulation algorithms like the hybrid molecular dynamics algorithm [13,14].

On the other hand, in the framework of the multi-boson algorithm [15], one can define an exact single flavor simulation scheme for Wilson fermions

---

[2] "Dynamical" in contrast to "quenched" simulations that neglect fermionic loops.
[3] The naive discretization of the Dirac operator yields, in addition to the true mode, 15 "doublers".

[16–18]. Here many additional bosonic auxiliary fields are required, and it is not clear if the method would be well suited for more complicated actions or allows to exploit sophisticated preconditioning.

In this paper, we propose to include the Wilson fermion determinant into HMC such that a single quark can be described. We show that the fermionic determinant can be written as a product of the determinants of two *hpd* matrices. This is achieved by the representation of the fermion matrix via its Schur complement. Both determinants can be treated by the hybrid Monte Carlo in the standard manner. There are, however, some caveats: one of the two matrices involves the inverse of the other one. As a consequence, a nested inversion must be carried out. We propose to use a Uzawa-like inverter which allows to solve the nested system within one iterative process. A second difficulty stems from the initialization of the fermion action in the HMC algorithm: since we have a single flavor system, we have to compute a square root at the beginning of a HMC trajectory.

The paper is organized as follows: in section 2 we introduce the basics of lattice QCD and will discuss the difficulties simulating single flavors of Wilson fermions by the HMC. In section 3 we give the transformation of the fermion matrix using the Schur complement, in section 4 we present the non-mass-degenerate HMC for Wilson fermions along with a discussion of numerical problems. In section 5, we introduce an algorithm that can compute the nested inversion in one iterative process. Finally, we summarize and give an outlook.

2   Numerical Problem

The action of lattice QCD is the sum of a gauge part (which is not relevant for the following) and a fermionic part:

$$S[U, \bar{\psi}, \psi] := S_g[U] + \sum_{x,y,a,b,\alpha,\beta} \bar{\psi}_x^a \, M_{x,y}^{a,b} \, \psi_y^b. \tag{1}$$

V is 4-dimensional volume in Euclidean space. The full matrix is a tensor product with 3×3 SU(3) matrices U (color) and four 4×4 matrices γ_μ (Dirac spin), hence

$$M \in \mathbb{C}^{12V \times 12V}. \tag{2}$$

Here the indices x an y stands for the 4-dimensional coordinates, a and b denote color space and α and β Dirac space indices.

The fermion fields ψ_x are *Grassmann* variables. Therefore, the Grassmann integral of the exponential of the fermionic part of the action, i.e., the fermionic part of the path integral, can be carried out,

$$\int [d\psi][d\bar{\psi}] \, e^{-\bar{\psi} M \psi} = \det(M), \tag{3}$$

to yield the determinant of M for arbitrary complex M.

Wilson fermions are defined by the interaction matrix

$$M = \mathbf{1} - \kappa D, \qquad M \in \mathbb{C}^{12V \times 12V},$$

$$D_{x,y} = \sum_{\mu=1}^{4} (\mathbf{1} - \gamma_\mu)\, U_\mu(x)\, \delta_{x,y-\mu} + (\mathbf{1} + \gamma_\mu)\, U_\mu^\dagger(x-\mu)\, \delta_{x,y+\mu}. \qquad (4)$$

The Dirac matrices in Euclidean space satisfy the anti-commutator relations

$$\gamma_\mu \gamma_\nu + \gamma_\nu \gamma_\mu = 2\delta_{\mu\nu}. \qquad (5)$$

It is important for the following considerations to work with the chiral representation of the Dirac matrices, as given in the appendix A. In this representation, γ_5, defined as the product $\gamma_5 = \gamma_4 \gamma_1 \gamma_2 \gamma_3$, is diagonal.

M is a complex non-hermitian matrix which moreover is not normal, i.e.,

$$MM^\dagger \neq M^\dagger M. \qquad (6)$$

Thus M cannot be diagonalized by a unitary transformation. At most, it might be diagonalizable by a similarity transformation. However, M exhibits the so-called γ_5-symmetry or γ_5-hermiticity:

$$M\gamma_5 = \gamma_5 M. \qquad (7)$$

Therefore, the matrix

$$Q := \gamma_5 M \qquad (8)$$

is hermitian. We can immediately read off the chiral representation of the Dirac matrices, see the appendix, that, for $\kappa = 0$, Q exhibits an equal number of positive and negative eigenvalues. In general, Q is indefinite.

The matrix $M^\dagger M$ is *hpd*. Since the determinant of $M^\dagger M$ represents two mass degenerate fermion flavors,

$$\det\left(M^\dagger M\right) = \det\left(M\right)\det\left(M\right), \qquad (9)$$

the two flavor situation is ideal for HMC simulations. Being unitary diagonalizable it can be represented by a Gaussian integral over complex fields ϕ:

$$\det\left(M\right)\det\left(M\right) = \left(\frac{1}{2\pi}\right)^{12V} \int [d\phi][d\phi^*] e^{-\phi^\dagger \left(M^\dagger M\right)^{-1}\phi}. \qquad (10)$$

For the single flavor M, such a simple construction fails even for the situation $\kappa < \kappa_c$ where M is positive definite, because the Jacobian of the similarity transform—if the diagonalization is feasible at all—is not equal to 1.

On the other hand,

$$\det(M) = \det(Q). \tag{11}$$

But, Q is indefinite. Therefore, a Gaussian integral representation for its determinant does not exist.

We conclude that so far there is no direct way to include individual flavors within the HMC algorithm.

3   Schur Complement of Q

Let's consider the hermitian matrix Q with the Dirac matrices given in the chiral representation of appendix A. The explicit form of Q in the chiral representation is given by

$$Q = \begin{pmatrix} Q_{11} & Q_{12} \\ Q_{21} & Q_{22} \end{pmatrix} = \begin{pmatrix} \mathbf{1}(1 - \kappa D_{11}) & -\kappa D_{12} \\ -\kappa D_{12}^\dagger & -\mathbf{1}(1 - \kappa D_{11}) \end{pmatrix}. \tag{12}$$

Here, the bold face $\mathbf{1}$ represents a unit matrix in 2×2 spin space. D_{11} does not carry spin indices while D_{12} consists of 2×2 blocks in spin space,

$$D_{11} = \sum_{\mu=1}^{4} U_\mu(\mathsf{x}) \, \delta_{\mathsf{x},\mathsf{x}+\mu} + U_\mu^\dagger(\mathsf{x}-\mu) \, \delta_{\mathsf{x},\mathsf{x}-\mu}$$

$$D_{12} = \sum_{\mu=1}^{4} \eta_\mu \left[U_\mu(\mathsf{x}) \, \delta_{\mathsf{x},\mathsf{x}+\mu} - U_\mu^\dagger(\mathsf{x}-\mu) \, \delta_{\mathsf{x},\mathsf{x}-\mu} \right], \tag{13}$$

with

$$\eta_1 = i\sigma_1, \eta_2 = i\sigma_2, \\ \eta_3 = i\sigma_3, \eta_4 = -\mathbf{1}. \tag{14}$$

The Schur complement of a 2×2 block matrix with non-singular block A is given by

$$\begin{bmatrix} A & B \\ C & D \end{bmatrix} = \begin{bmatrix} \mathbf{1} & 0 \\ CA^{-1} & \mathbf{1} \end{bmatrix} \begin{bmatrix} A & 0 \\ 0 & D - CA^{-1}B \end{bmatrix} \begin{bmatrix} \mathbf{1} & A^{-1}B \\ 0 & \mathbf{1} \end{bmatrix}. \tag{15}$$

Hence

$$\det(Q) = \det\left(\mathbf{1}(1 - \kappa D_{11})\right) \det\left(\mathbf{1}(1 - \kappa D_{11}) + \kappa^2 D_{12}[\mathbf{1}(1 - \kappa D_{11})]^{-1} D_{12}^\dagger\right). \tag{16}$$

The minus sign of the Q_{22} term cancels as the rank of the matrix is even. From Eq. 13 we now that the first matrix is diagonal in Dirac space. Therefore its determinant can be written as follows:

$$\det(Q) = \det\left(\mathbf{1}(1 - \kappa D_{11})\right) = [\det(1 - \kappa D_{11})]^2 = \det(1 - \kappa D_{11})^2, \quad (17)$$

where $Q_w^2 = (1 - D_{11})^2$ is hpd for any κ. Q_w carries no spin index.

The second matrix

$$Q_{sc} = \mathbf{1}(1 - \kappa D_{11}) + \kappa^2 D_{12}[\mathbf{1}(1 - \kappa D_{11})]^{-1} D_{12}^{\dagger}, \quad (18)$$

the Schur complement, is hermitian, as it is the sum of hermitian matrices. For $\kappa = 0$, it is hpd, thus there exists some value of κ, κ_c^{sc}, for which it becomes indefinite. For $\kappa < \kappa_c^{sc}$, Q_{sc} is hpd.

4 One Flavor HMC

We have shown that

$$\det\left(M\right) = \det\left(Q\right) = \det\left(Q_w\right)\det\left(Q_{sc}\right), \quad (19)$$

for Q_{11} being non-singular. If $\kappa < \kappa_{sc}$, both matrices are hpd and can be represented by Gaussian integrals. Let $\phi \in \mathbb{C}^{3V}$ and $\chi \in \mathbb{C}^{6V}$:

$$\det\left(Q_w\right)\det\left(Q_{sc}\right)$$
$$= \left(\frac{1}{2\pi}\right)^{3V}\left(\frac{1}{2\pi}\right)^{6V}\int [d\phi][d\phi^*][d\chi][d\chi^*]e^{-\phi^{\dagger}(Q_w)^{-2}\phi - \chi^{\dagger}(Q_{sc})^{-1}\chi}. \quad (20)$$

The fermionic part of the path integral being expressed in terms of the pseudo-fermion degrees of freedom ϕ and χ, the HMC algorithm can proceed as usual [19], see the Φ-algorithm in Ref. [6]. The heat bath for the ϕ fields is trivial like in the previous case of mass-degenerate QCD. Let $R \in \mathbb{C}^{3V}$ be Gaussian distributed. With

$$\phi = Q_w^{\dagger} R, \quad (21)$$

$$R^{\dagger} R = \phi^{\dagger}(Q_w)^{-2}\phi. \quad (22)$$

and

$$(2\pi)^{3V}\det(Q_w^2) = \int [d\phi][d\phi^*]e^{-\phi^{\dagger}(Q_w)^{-2}\phi}. \quad (23)$$

The refreshment of the χ fields causes more trouble, however. In order to generate the required distribution, we have to solve a linear system involving the square root of Q_{sc}. Let $R \in \mathbb{C}^{6V}$ be Gaussian distributed:

$$\chi = Q_{sc}^{\frac{1}{2}} R. \quad (24)$$

Then

$$R^\dagger R = \chi^\dagger (Q_{sc})^{-1} \chi \tag{25}$$

and

$$(2\pi)^{6V} \det(Q_{sc}) = \int [d\chi][d\chi^*] e^{-\chi^\dagger (Q_{sc})^{-1}\chi} \tag{26}$$

Even though the solution of Eq. 25 is only required at the beginning of a trajectory, it is particularly expensive as we must cope with the internal inversion of Q_{11}. The implications of this issue are not clarified yet. As an alternative, we might include the dynamics of the χ fields into HMC avoiding the expensive heat bath step for Q_{sc}, similar to the R-algorithm of Ref. [6].

5   Inexact Uzawa Iterations

The nested inversion within the solution of

$$Q_{sc} X = \chi \tag{27}$$

has to be carried out at each molecular dynamics time step of HMC as well as in the computation of the action. By Uzawa iterations [20] we avoid the solution of the linear system represented by Q_w within each iteration step of the Q_{sc} iteration.

The problem to solve is given by:

$$(Q_w + D_{12} Q_w^{-1} D_{12}^\dagger) X = \chi, \tag{28}$$

with $X, \psi \in \mathbb{C}^{6V}$.

Let's consider the Jacobi iteration for Q_{sc},

$$X^{(i+1)} = \chi + \kappa D_{11} X^{(i)} - D_{12} Q_w^{-1} D_{12}^\dagger X^{(i)}. \tag{29}$$

The fixed-point solution X of Eq. 29, defined by

$$||X^{(i+1)} - X^{(i)}|| < \epsilon, \quad \epsilon \to 0 \tag{30}$$

solves equation Eq. 28, as can be easily verified.

Next we transform the simple Jacobi iteration into the Uzawa iteration:

$$\begin{aligned} X^{(i+1)} &= \chi + \kappa D_{11} X^{(i)} - D_{12} Y^{(i)} \\ Y^{(i+1)} &= Q_w^{-1} D_{12}^\dagger X^{(i)}. \end{aligned} \tag{31}$$

At this stage, the Uzawa scheme would certainly not offer an advantage as still we need to carry out an inversion in each iteration step. Let's turn towards the inexact Uzawa iteration. Choose

$$P_w \approx Q_w \tag{32}$$

as a preconditioner for Q_w which is easily invertible.

$$X^{(i+1)} = \chi + \kappa D_{11} X^{(i)} - D_{12} Y^{(i)}$$
$$Y^{(i+1)} = Y^{(i)} + P_w^{-1}(D_{12}^{\dagger} X^{(i)} - Q_w Y^{(i)}) \tag{33}$$

The fixed-point solution of Eq. 33, defined by

$$||X^{(i+1)} - X^{(i)}|| < \epsilon,$$
$$||Y^{(i+1)} - Y^{(i)}|| < \epsilon'. \tag{34}$$

for $\epsilon, \epsilon' \to 0$, solves Eq. 28. Using the second relation, one finds

$$0 = P_w^{-1}(-Q_w Y + D_{12}^{\dagger} X),$$
$$\Leftrightarrow Y = Q_w^{-1} D_{12}^{\dagger} X. \tag{35}$$

As a cheap preconditioner we can use a truncated Neumann series for Q_w^{-1}:

$$P_w^{-1} = (1 - \kappa D_{11})^{-1} = 1 + \kappa D_{11} + \dots \tag{36}$$

One has to find the optimal length of this series compared to the convergence of the outer iteration. In our implementation we have chosen the simple first-order approximation.

$$P_w^{-1} = 1 + \kappa D_{11}. \tag{37}$$

We have made first tests of the Uzawa iteration in comparison to a conjugate gradient (CG) for the inversion of Eq. 28. We used a quenched 16^4 configuration at $\beta = 6.0$ at a moderate κ-value of $\kappa = 0.12$. The outer CG took about 240 iterations while the inner iteration took about 50 steps. The Uzawa iteration required about 610 steps in comparison. The overall improvement for this setting is about a factor of 10. Of course, these results are very preliminary and have to be confirmed for other values of κ and β with realistic HMC simulations. It would also be desirable to implement the Uzawa scheme within Krylov subspace solvers.

For free theory ($U = 1$), one can diagonalize D_{11} by Fourier transformation. As in the case of Wilson fermions, one finds $\kappa_c = \frac{1}{8}$. We have, so far, investigated the lowest lying eigenvalues of Q_w for the interacting case: on a $16^3 \times 32$ configuration, taken from the SESAM ensembles at $\beta = 5.6$ and $\kappa_{\text{sea}} = 0.1575$, we found for the inverse of the critical eigenvalue of D_{11}, $1/\lambda_{min} = 0.144$. The critical κ of D_{11} is reached for a *kappa*-value being smaller than the critical κ value of the Wilson matrix. Therefore, the singularity might be a barrier screening the approach to the chiral limit. These questions have to be clarified in an actual HMC simulation with one-flavor QCD.

6   Summary and Outlook

The determinant of the Wilson fermion matrix is equivalent to the product of the determinants of two hermitian positive definite matrices. This formulation allows to simulate quark flavors with individual mass values by means of the HMC algorithm. The solution of the ensuing linear system is hampered by a nested inversion problem. However, it can be treated by use of Uzawa iterations.

Currently we perform a feasibility study to gain first experiences with the new method. In particular, we are interested in the analytic structure of Q_w and Q_{sc}. Furthermore, we construct an efficient heat-bath for the refreshment of the pseudo-fermions for Q_{sc} based on the polynomial representation of the square root of a truncated polynomial that represents Q_{sc}.

We have presented the method for standard Wilson fermions. The Schur decomposition can as well be applied to improved Wilson fermions with clover term. Work on this line is in progress.

Acknowledgments

I thank B. Bunk for important discussions that motivated the present work. I am happy to work with A. Frommer, B Medeke, and K. Schilling from Wuppertal university who contributed with interesting ideas. In particular I thank H. Neff who has kindly performed the eigenvalue computations. I am grateful to M. Peardon and Ph. de Forcrand who pointed out important simplifications of the method during my talk at the workshop.

A   Chiral Dirac Matrices

Euclidean Dirac matrices in the chiral representation:

$$
\begin{aligned}
\gamma_1 &= \begin{pmatrix} 0 & 0 & 0 & -i \\ 0 & 0 & -i & 0 \\ 0 & i & 0 & 0 \\ i & 0 & 0 & 0 \end{pmatrix} = i \begin{pmatrix} 0 & -\sigma_1 \\ \sigma_1 & 0 \end{pmatrix} \\[6pt]
\gamma_2 &= \begin{pmatrix} 0 & 0 & 0 & -1 \\ 0 & 0 & 1 & 0 \\ 0 & 1 & 0 & 0 \\ -1 & 0 & 0 & 0 \end{pmatrix} = i \begin{pmatrix} 0 & -\sigma_2 \\ \sigma_2 & 0 \end{pmatrix} \\[6pt]
\gamma_3 &= \begin{pmatrix} 0 & 0 & -i & 0 \\ 0 & 0 & 0 & i \\ i & 0 & 0 & 0 \\ 0 & -i & 0 & 0 \end{pmatrix} = i \begin{pmatrix} 0 & -\sigma_3 \\ \sigma_3 & 0 \end{pmatrix} \\[6pt]
\gamma_4 &= \begin{pmatrix} 0 & 0 & 1 & 0 \\ 0 & 0 & 0 & 1 \\ 1 & 0 & 0 & 0 \\ 0 & 1 & 0 & 0 \end{pmatrix} = \begin{pmatrix} 0 & 1 \\ 1 & 0 \end{pmatrix} \\[6pt]
\gamma_5 &= \begin{pmatrix} 1 & 0 & 0 & 0 \\ 0 & 1 & 0 & 0 \\ 0 & 0 & -1 & 0 \\ 0 & 0 & 0 & -1 \end{pmatrix}
\end{aligned}
\tag{38}
$$

References

1. C. Caso et al. (Particle Data Group). Review of particle properties. *European Phys. Jour.*, C3, 1999.
2. J. Gasser and H. Leutwyler. Quark masses. *Phys. Rept.*, 87:77–169, 1982.
3. A. Frommer, V. Hannemann, B. Nöckel, Th. Lippert, and K. Schilling. Accelerating Wilson fermion matrix inversions by means of the stabilized biconjugate gradient algorithm. *Int. J. Mod. Phys.*, C5:1073, 1994.
4. S. Fischer, A. Frommer, U. Glässner, Th. Lippert, G. Ritzenhöfer, and K. Schilling. A parallel SSOR preconditioner for lattice QCD. *Comp. Phys. Commun.*, 98:20–34, 1996.
5. S. Duane, A. D. Kennedy, B. J. Pendleton, and D. Roweth. Hybrid Monte Carlo. *Phys. Lett.*, B195:216, 1987.
6. S. Gottlieb, W. Liu, D. Toussaint, R. L. Renken, and R. L. Sugar. Hybrid molecular dynamics algorithms for the numerical simulation of quantum chromodynamics. *Phys. Rev.*, D35:2531, 1987.
7. D. H. Weingarten and D. N. Petcher. Monte Carlo integration for lattice gauge theories with fermions. *Phys. Lett.*, B99:333, 1981.
8. K. G. Wilson. Quarks: from paradox to myth. In *New Phenomena In Subnuclear Physics*, volume Part A, pages 13–22. New York, 1977, 1975. Proceedings of Erice summer school, Erice.

9. H. Neuberger. The overlap Dirac operator. In Frommer et al. [21]. These proceedings.

10. S. Sharpe and N. Shoresh. Partially quenched QCD with non-degenerate dynamical quarks. 1999.

11. N. Eicker, P. Lacock, K. Schilling, A. Spitz, U. Glässner, S. Güsken, H. Hoeber, Th. Lippert, Th. Struckmann, P. Ueberholz, and J. Viehoff. Light and strange hadron spectroscopy with dynamical Wilson fermions. *Phys. Rev.*, D59:014509, 1999.

12. A. Ali Khan et al. Light hadron spectrum and quark masses in QCD with two flavors of dynamical quarks. hep-lat/9909050, 1999.

13. Claude Bernard et al. The static quark potential in three flavor qcd. 2000.

14. A. Peikert, F. Karsch, E. Laermann, and B. Sturm. The three flavour chiral phase transition with an improved quark and gluon action in lattice QCD. *Nucl. Phys. Proc. Suppl.*, 73:468, 1999.

15. M. Lüscher. A new approach to the problem of dynamical quarks in numerical simulations of lattice QCD. *Nucl. Phys.*, B418:637–648, 1994.

16. C. Alexandrou et al. The deconfinement phase transition in one-flavor QCD. *Phys. Rev.*, D60:034504, 1999.

17. C. Alexandrou et al. Thermodynamics of one-flavour QCD. Preprint hep-lat/9806004, 1998.

18. I. Montvay. Least-squares optimized polynomials for fermion simulations. In Frommer et al. [21]. These proceedings.

19. Th. Lippert. The hybrid Monte Carlo algorithm for quantum chromodynamics. In H. Meyer-Ortmanns and A. Klümper, editors, *Field theoretical tools for polymer and particle physics*, volume 508 of *Lecture Notes in Physics*, pages 122–131. Springer Verlag, Heidelberg, 1998. Proceedings of Workshop of the Graduiertenkolleg at the University of Wuppertal, June, 1997.

20. J. H. Bramble, J. E. Pasciak, and A. T. Vassilev. Analysis of the inexact Uzawa algorithm for saddle point problems. *SIAM J. Numer. Anal.*, 34:1072–1092, 1997.

21. A. Frommer, Th. Lippert, B. Medeke, and K. Schilling, editors. *Numerical Challenges in Lattice QCD*. Springer Verlag, Heidelberg, 2000. Proceedings of International Workshop, University of Wuppertal, August 22-24, 1999, Present proceedings.

Author Index

Subject Index

Editorial Policy

§1. Volumes in the following four categories will be published in LNCSE:

i)   Research monographs
ii) Lecture and seminar notes
iii) Conference proceedings
iv) Textbooks

Those considering a book which might be suitable for the series are strongly advised to contact the publisher or the series editors at an early stage.

§2. Categories i) and ii). These categories will be emphasized by Lecture Notes in Computational Science and Engineering. **Submissions by interdisciplinary teams of authors are encouraged.** The goal is to report new developments – quickly, informally, and in a way that will make them accessible to non-specialists. In the evaluation of submissions timeliness of the work is an important criterion. Texts should be well-rounded, well-written and reasonably self-contained. In most cases the work will contain results of others as well as those of the author(s). In each case the author(s) should provide sufficient motivation, examples, and applications. In this respect, Ph.D. theses will usually be deemed unsuitable for the Lecture Notes series. Proposals for volumes in these categories should be submitted either to one of the series editors or to Springer-Verlag, Heidelberg, and will be refereed. A provisional judgment on the acceptability of a project can be based on partial information about the work: a detailed outline describing the contents of each chapter, the estimated length, a bibliography, and one or two sample chapters – or a first draft. A final decision whether to accept will rest on an evaluation of the completed work which should include

– at least 100 pages of text;
– a table of contents;
– an informative introduction perhaps with some historical remarks which should be
  accessible to readers unfamiliar with the topic treated;
– a subject index.

§3. Category iii). Conference proceedings will be considered for publication provided that they are both of exceptional interest and devoted to a single topic. One (or more) expert participants will act as the scientific editor(s) of the volume. They select the papers which are suitable for inclusion and have them individually refereed as for a journal. Papers not closely related to the central topic are to be excluded. Organizers should contact Lecture Notes in Computational Science and Engineering at the planning stage.

In exceptional cases some other multi-author-volumes may be considered in this category.

§4. Category iv) Textbooks on topics in the field of computational science and engineering will be considered. They should be written for courses in CSE education. Both graduate and undergraduate level are appropriate. Multidisciplinary topics are especially welcome.

§5. Format. Only works in English are considered. They should be submitted in camera-ready form according to Springer-Verlag's specifications. Electronic material can be included if appropriate. Please contact the publisher. Technical instructions and/or TEX macros are available via http://www.springer.de/author/tex/help-tex.html; the name of the macro package is "LNCSE – LaTEX2e class for Lecture Notes in Computational Science and Engineering". The macros can also be sent on request.

General Remarks

Lecture Notes are printed by photo-offset from the master-copy delivered in camera-ready form by the authors. For this purpose Springer-Verlag provides technical instructions for the preparation of manuscripts. See also *Editorial Policy*.

Careful preparation of manuscripts will help keep production time short and ensure a satisfactory appearance of the finished book. The actual production of a Lecture Notes volume normally takes approximately 12 weeks.

The following terms and conditions hold:

Categories i), ii), and iii):
Authors receive 50 free copies of their book. No royalty is paid. Commitment to publish is made by letter of intent rather than by signing a formal contract. Springer-Verlag secures the copyright for each volume.

For conference proceedings, editors receive a total of 50 free copies of their volume for distribution to the contributing authors.

Category iv):
Regarding free copies and royalties, the standard terms for Springer mathematics monographs and textbooks hold. Please write to Peters@springer.de for details. The standard contracts are used for publishing agreements.

All categories:
Authors are entitled to purchase further copies of their book and other Springer mathematics books for their personal use, at a discount of 33,3 % directly from Springer-Verlag.

Addresses:

Professor M. Griebel
Institut für Angewandte Mathematik
der Universität Bonn
Wegelerstr. 6
D-53115 Bonn, Germany
e-mail: griebel@iam.uni-bonn.de

Professor D. E. Keyes
Computer Science Department
Old Dominion University
Norfolk, VA 23529–0162, USA
e-mail: keyes@cs.odu.edu

Professor R. M. Nieminen
Laboratory of Physics
Helsinki University of Technology
02150 Espoo, Finland
e-mail: rni@fyslab.hut.fi

Professor D. Roose
Department of Computer Science
Katholieke Universiteit Leuven
Celestijnenlaan 200A
3001 Leuven-Heverlee, Belgium
e-mail: dirk.roose@cs.kuleuven.ac.be

Professor T. Schlick
Department of Chemistry and
Courant Institute of Mathematical
Sciences
New York University
and Howard Hughes Medical Institute
251 Mercer Street, Rm 509
New York, NY 10012-1548, USA
e-mail: schlick@nyu.edu

Springer-Verlag, Mathematics Editorial
Tiergartenstrasse 17
D-69121 Heidelberg, Germany
Tel.: *49 (6221) 487-185
e-mail: peters@springer.de
http://www.springer.de/math/
peters.html

Lecture Notes in Computational Science and Engineering

Vol. 1 D. Funaro, *Spectral Elements for Transport-Dominated Equations.* 1997. X, 211 pp. Softcover. ISBN 3-540-62649-2

Vol. 2 H. P. Langtangen, *Computational Partial Differential Equations.* Numerical Methods and Diffpack Programming. 1999. XXIII, 682 pp. Hardcover. ISBN 3-540-65274-4

Vol. 3 W. Hackbusch, G. Wittum (eds.), *Multigrid Methods V.* Proceedings of the Fifth European Multigrid Conference held in Stuttgart, Germany, October 1-4, 1996. 1998. VIII, 334 pp. Softcover. ISBN 3-540-63133-X

Vol. 4 P. Deuflhard, J. Hermans, B. Leimkuhler, A. E. Mark, S. Reich, R. D. Skeel (eds.), *Computational Molecular Dynamics: Challenges, Methods, Ideas.* Proceedings of the 2nd International Symposium on Algorithms for Macromolecular Modelling, Berlin, May 21-24, 1997. 1998. XI, 489 pp. Softcover. ISBN 3-540-63242-5

Vol. 5 D. Kröner, M. Ohlberger, C. Rohde (eds.), *An Introduction to Recent Developments in Theory and Numerics for Conservation Laws.* Proceedings of the International School on Theory and Numerics for Conservation Laws, Freiburg / Littenweiler, October 20-24, 1997. 1998. VII, 285 pp. Softcover. ISBN 3-540-65081-4

Vol. 6 S. Turek, *Efficient Solvers for Incompressible Flow Problems.* An Algorithmic and Computational Approach. 1999. XVII, 352 pp, with CD-ROM. Hardcover. ISBN 3-540-65433-X

Vol. 7 R. von Schwerin, *Multi Body System SIMulation.* Numerical Methods, Algorithms, and Software. 1999. XX, 338 pp. Softcover. ISBN 3-540-65662-6

Vol. 8 H.-J. Bungartz, F. Durst, C. Zenger (eds.), *High Performance Scientific and Engineering Computing.* Proceedings of the International FORTWIHR Conference on HPSEC, Munich, March 16-18, 1998. 1999. X, 471 pp. Softcover. 3-540-65730-4

Vol. 9 T. J. Barth, H. Deconinck (eds.), *High-Order Methods for Computational Physics.* 1999. VII, 582 pp. Hardcover. 3-540-65893-9

Vol. 10 H. P. Langtangen, A. M. Bruaset, E. Quak (eds.), *Advances in Software Tools for Scientific Computing.* 2000. X, 357 pp. Softcover. 3-540-66557-9

Vol. 11 B. Cockburn, G. E. Karniadakis, C.-W. Shu (eds.), *Discontinuous Galerkin Methods.* Theory, Computation and Applications. 2000. XI, 470 pp. Hardcover. 3-540-66787-3

Vol. 12 U. van Rienen, *Numerical Methods in Computational Electrodynamics.* Linear Systems in Practical Applications. 2000. XIII, 375 pp. Softcover. 3-540-67629-5

Vol. 13 B. Engquist, L. Johnsson, M. Hammill, F. Short (eds.), *Simulation and Visualization on the Grid.* Parallelldatorcentrum Seventh Annual Conference, Stockholm, December 1999, Proceedings. 2000. XIII, 301 pp. Softcover. 3-540-67264-8

Vol. 14 E. Dick, K. Riemslagh, J. Vierendeels (eds.), *Multigrid Methods VI.* Proceedings of the Sixth European Multigrid Conference Held in Gent, Belgium, September 27-30, 1999. 2000. IX, 293 pp. Softcover. 3-540-67157-9

Vol. 15 A. Frommer, T. Lippert, B. Medeke, K. Schilling (eds.), *Numerical Challenges in Lattice Quantum Chromodynamics.* Joint Interdisciplinary Workshop of John von Neumann Institute for Computing, Jülich and Institute of Applied Computer Science, Wuppertal University, August 1999. 2000. VIII, 184 pp. Softcover. 3-540-67732-1

Vol. 16 J. Lang, *Adaptive Multilevel Solution on Nonlinear Parabolic PDE Systems.* Theory, Algorithm and Applications. 2000. VIII, 153 pp. Softcover. 3-540-67900-6

For further information on these books please have a look at our mathematics catalogue at the following URL: http://www.springer.de/math/index.html